Infectious Diseases and Substance Abuse

INFECTIOUS AGENTS AND PATHOGENESIS

Series Editors: Mauro Bendinelli, *University of Pisa*
Herman Friedman, *University of South Florida*
College of Medicine

Recent volumes in this series:

INFECTIOUS DISEASES AND SUBSTANCE ABUSE
Edited by Herman Friedman, Thomas W. Klein, and Mauro Bendinelli

CHLAMYDIA PNEUMONIAE
Infection and Disease
Edited by Herman Friedman, Yoshimasa Yamamoto,
and Mauro Bendinelli

DNA TUMOR VIRUSES
Oncogenic Mechanisms
Edited by Giuseppe Barbanti-Brodano, Mauro Bendinelli,
and Herman Friedman

ENTERIC INFECTIONS AND IMMUNITY
Edited by Lois J. Paradise, Mauro Bendinelli,
and Herman Friedman

HELICOBACTER PYLORI INFECTION AND IMMUNITY
Edited by Yoshimasa Yamamoto, Herman Friedman,
and Paul S. Hoffman

HERPESVIRUSES AND IMMUNITY
Edited by Peter G. Medveczky, Herman Friedman,
and Mauro Bendinelli

HUMAN RETROVIRAL INFECTIONS
Immunological and Therapeutic Control
Edited by Kenneth E. Ugen, Mauro Bendinelli,
and Herman Friedman

MICROORGANISMS AND AUTOIMMUNE DISEASES
Edited by Herman Friedman, Noel R. Rose,
and Mauro Bendinelli

OPPORTUNISTIC INTRACELLULAR BACTERIA AND IMMUNITY
Edited by Lois J. Paradise, Herman Friedman,
and Mauro Bendinelli

PULMONARY INFECTIONS AND IMMUNITY
Edited by Herman Chmel, Mauro Bendinelli,
and Herman Friedman

RAPID DETECTION OF INFECTIOUS AGENTS
Edited by Steven Specter, Mauro Bendinelli,
and Herman Friedman

A Continuation Order Plan is available for this series. A continuation order will bring delivery of each new volume immediately upon publication. Volumes are billed only upon actual shipment. For further information, please contact the publisher.

Infectious Diseases and Substance Abuse

Edited by

Herman Friedman,
Thomas W. Klein
Department of Medical Microbiology and Immunology,
University of South Florida College of Medicine, Tampa, FL

and

Mauro Bendinelli
Department of Experimental Pathology,
University of Pisa, Pisa, Italy

A C.I.P. record for this book is available from the Library of Congress

ISBN-10: 0-306-48687-3   e-ISBN: 0-306-48688-1
ISBN-13: 978-0306-48687-6

Printed on acid-free paper

Printed in the United States of America.   (NEW / EB)

9 8 7 6 5 4 3 2 1

springeronline.com

Contributors

NORMA C. ALONZO • Departments of Pharmacology and Neuroscience, Georgetown University Medical Center, Washington, DC 20007

ALBERT H. AVILA • Departments of Pharmacology and Neuroscience, Georgetown University Medical Center, Washington, DC 20007

GREGORY J. BAGBY • Department of Medicine, Section of Pulmonary and Critical Care Medicine, Department of Physiology, Alcohol Research Center, Louisiana State University Health Science Center, New Orleans, LA 70112

GAYLE C. BALDWIN • Division of Hematology/Oncology, Department of Medicine, David Geffen School of Medicine at UCLA, Los Angeles, CA 90095

RODERICK A. BARKE • Departments of Pharmacology and Surgery, University of Minnesota, Minneapolis, MN 55455

BARBARA M. BAYER • Departments of Pharmacology and Neuroscience, Georgetown University Medical Center, Washington, DC 20007

FILIP BEDNAR • Center for Substance Abuse Research, Fels Institute for Cancer Research and Molecular Biology, Temple University School of Medicine, Philadelphia, PA 19140

JERRY L. BULEN • Department of Medical Microbiology and Immunology, University of South Florida, College of Medicine, Tampa, FL 33612-4799

GUY A. CABRAL • Department of Microbiology and Immunology, Virginia Commonwealth University, Richmond, VA 23298-0678

FRANCINE MARCIANO-CABRAL • Department of Microbiology and Immunology, Virginia Commonwealth University, Richmond, VA 23298-0678

PENELOPE C. DAVEY • Center for Substance Abuse Research, Fels Institute for Cancer Research and Molecular Biology, Temple University School of Medicine, Philadelphia, PA 19140

HERMAN FRIEDMAN • Department of Medical Microbiology and Immunology, University of South Florida, College of Medicine, Tampa, FL 33612

RICARDO GOMEZ • Universidad Autónoma de Nuevo León, San Nicolás de los Garza, NL, Mexico

DAVID E. KAMINSKY • Center for Substance Abuse Research, Fels Institute for Cancer Research and Molecular Biology, Temple University School of Medicine, Philadelphia, PA 19140

THOMAS W. KLEIN • Department of Medical Microbiology and Immunology, University of South Florida, College of Medicine, Tampa, FL 33612-4799

JAY K. KOLLS • Department of Medicine, Section of Pulmonary and Critical Care Medicine, Alcohol Research Center and Gene Therapy Programs, Louisiana State University Health Science Center, New Orleans, LA 70112

KATHY McALLEN • Department of Pharmacology, University of Tennessee Health Science Center, Memphis, TN 38120

PETER G. MEDVECZKY • Department of Medical Microbiology and Immunology, University of South Florida, College of Medicine, Tampa, FL 33612-4799

STEVE NELSON • Department of Medicine, Section of Pulmonary and Critical Care Medicine, Department of Physiology, Alcohol Research Center, Louisiana State University Health Science Center, New Orleans, LA 70112

CATHERINE NEWTON • Department of Medical Microbiology and Immunology, University of South Florida, College of Medicine, Tampa, FL 33612

SUSAN PROSS • Department of Medical Microbiology and Immunology, University of South Florida, College of Medicine, Tampa, FL 33612

LEE J. QUINTON • Department of Physiology, Alcohol Research Center, Louisiana State University Health Science Center, New Orleans, LA 70112

SEDDIGHEH RAZANI-BOROUJERDI • Respiratory Immunology, Lovelace Respiratory Research Institute, Albuquerque, NM 87108

THOMAS J. ROGERS • Center for Substance Abuse Research, Fels Institute for Cancer Research and Molecular Biology, Temple University School of Medicine, Philadelphia, PA 19140

MICHAEL D. ROTH • Division of Pulmonary and Critical Care, Department of Medicine, David Geffen School of Medicine at UCLA, Los Angeles, CA 90095-1690.

SABITA ROY • Departments of Pharmacology and Surgery, University of Minnesota, Minneapolis, MN 55455

NAHID A. SHAHABI • Department of Pharmacology, University of Tennessee Health Science Center, Memphis, TN 38120

BURT M. SHARP • Department of Pharmacology, University of Tennessee Health Science Center, Memphis, TN 38120

SHASHI P. SINGH • Respiratory Immunology, Lovelace Respiratory Research Institute, Albuquerque, NM 87108

MOHAN L. SOPORI • Respiratory Immunology, Lovelace Respiratory Research Institute, Albuquerque, NM 87108

AMBER D. STEELE • Center for Substance Abuse Research, Fels Institute for Cancer Research and Molecular Biology, Temple University School of Medicine, Philadelphia, PA 19140

JING-HUA WANG • Departments of Pharmacology and Surgery, University of Minnesota, Minneapolis, MN 55455

RICHARD WEBER • Department of Biomedical and Therapeutic Sciences, University of Illinois College of Medicine at Peoria, Peoria, IL

YOSHIMASA YAMAMOTO • Department of Basic Laboratory Sciences, Osaka University Graduate School of Medicine, Osaka, JAPAN

ZEKI YUMUK • Department of Clinical Microbiology, Kocaeli University Faculty of Medicine, Kocaeli, TURKEY

PING ZHANG • Department of Medicine, Section of Pulmonary and Critical Care Medicine, Alcohol Research Center, Louisiana State University Health Science Center, New Orleans, LA 70112

Preface to the Series

The mechanisms of disease production by infectious agents are presently the focus of an unprecedented flowering of studies. The field has undoubtedly received impetus from the considerable advances recently made in the understanding of the structure, biochemistry, and biology of viruses, bacteria, fungi, and other parasites. Another contributing factor is our improved knowledge of immune responses and other adaptive or constitutive mechanisms by which hosts react to infection. Furthermore, recombinant DNA technology, monoclonal antibodies, and other newer methodologies have provided the technical tools for examining questions previously considered too complex to be successfully tackled. The most important incentive of all is probably the regenerated idea that infection might be the initiating event in many clinical entities presently classified as idiopathic or of uncertain origin.

Infectious pathogenesis research holds great promise. As more information is uncovered, it is becoming increasingly apparent that our present knowledge of the pathogenic potential of agents is often limited to the most noticeable effects, which sometimes represent only the tip of the iceberg. For example, it is now well appreciated that pathologic processes caused by infectious agents may emerge clinically after an incubation of decades and may result from genetic, immunologic, and other indirect routes more than from the infecting agent in itself. Thus, there is a general expectation that continued investigation will lead to the isolation of new agents of infection, the identification of hitherto unsuspected etiologic correlations, and, eventually, more effective approaches to prevention and therapy.

Studies on the mechanisms of disease caused by infectious agents demand a breadth of understanding across many specialized areas, as well as much cooperation between clinicians and experimentalists. The series *Infectious Agents and Pathogenesis* is intended not only to document the state of the art in this fascinating and challenging field but also to help lay bridges among diverse areas and people.

Mauro Bendinelli
Herman Friedman

Foreword and Introduction

The use of recreational drugs of abuse by large numbers of individuals in this country and abroad has aroused serious concerns about the consequences of this activity. For example, it is recognized that marijuana is currently widely used as a recreational drug in the United States as well as other countries. Similarly, abuse of cocaine, especially crack cocaine, is considered to be an epidemic. "The war on drugs" by the US Government was directly aimed at the illicit use of cocaine, marijuana, and opiates as well as other drugs of abuse. Furthermore, alcohol is also considered a major problem of abuse in this country as well as in many other countries. It is estimated there are at least 10 million alcoholics in the United States alone. A significant portion of those hospitalized with infectious diseases are alcoholics. Similarly, there have been many reports of association between marijuana use and increased susceptibility to infection as well as a relation between use of opiates and infections. The relationship between drug abuse and increased incidence of various infections has stimulated increased investigation of whether and how such drugs affect immune function, especially important for resistance against infectious agents.

During the last decades, a wide variety of studies have shown that drugs of abuse, including marijuana, cocaine, or opiates, as well as alcohol, alter both neurophysiological as well as pathological responses of individuals. Similarly, it has been shown that illicit drug use also alters immune function, and the influence of such altered immunity has marked physiological and physical consequences on drug abusers. Specifically, data have accumulated indicating that drugs of abuse markedly affect the immune response in both human populations and in experimental animal models, both *in vivo* and *in vitro*.

Experimental studies concerning microbial infections in animals have supported empirical observations reported earlier that many drugs of abuse are often associated with enhanced susceptibility to infectious diseases. Furthermore, the mechanisms whereby such drugs increase the likelihood of infections in humans as well as experimental animals have begun to be delineated. In particular, it has been shown that morphine, cocaine, or marijuana, as well as alcohol, enhance susceptibility to infection by bacteria, viruses, protozoa, or fungi when given to experimental animals or used to treat lymphoid cell populations *in vitro*. The purpose of this volume is to focus attention on valuable new information concerning the effects of recreational drugs on modulation of immune responses, especially pertaining to mechanisms important in resistance to infectious diseases, as well as to malignancy and autoimmunity. Studies concerning how illicit drugs affect immunity are considered even more urgent at the present time because of the worldwide epidemic of acquired immune deficiency syndrome (AIDS) caused by the human immunodeficiency virus (HIV). Infection with HIV causes the collapse of the immune system, making an individual highly susceptible to opportunistic microorganisms which cause significant clinical disease in mainly immunocompromised individuals.

The onset of the AIDS epidemic in the United States, and indeed worldwide, stimulated attempts to search for possible cofactors which result in a more rapid progression of the disease in individuals infected with HIV. Approximately a third of all AIDS patients in the United States and other developed countries are i.v. drug abusers. It has been shown that HIV is readily spread by contaminated needles or equipment used by drug abusers. However, it is also widely accepted that many illicit drugs not taken by the i.v. route but by other routes are immunosuppressive and modulate the immune system, especially by activation hypothalamic–pituitary–adrenal axis. Although many AIDS patients, especially in third world countries, are known not to be i.v. drug abusers, they often utilize drugs such as marijuana, cocaine, or even alcohol, and HIV may be transmitted by the sexual route, even in such drug abusers. Thus, there is much concern that such illicit drugs serve as a possible cofactor in the progression of AIDS.

There had been various studies during the past few years examining in detail the mechanisms whereby drugs of abuse compromise the immune system in general and specifically enhance susceptibility to infection. Thus, this book in the series *Infectious Agents and Pathogenesis* focuses specifically on possible relationships between drugs of abuse like cocaine, marijuana, and opiates, as well as alcohol, immune response function, and alteration of resistance to microorganisms, especially opportunistic bacteria. This volume presents a number of reviews concerning various categories of drugs, the immune system, and infectious disease. The first chapter is a detailed review by investigators from Georgetown University concerning the effects of both cocaine and morphine in animal models with regard to the nature and mechanism of immunomodulation resulting from acute withdrawal. The next chapter is by Drs. Baldwin and Roth from UCLA concerning links between cannabinoid use and HIV infection. Drs. Bulen and Medveczky from the University of South Florida then discuss the effects of cannabinoids on Herpesvirus infection and the mechanisms involved.

Drs. Guy Cabral and Francine Marciano-Cabral from the Medical College of Virginia describe studies concerning the effects of cannabinoids on increased susceptibility of brain cells to infection by an important amoeba known to cause neurologic disease. Investigations concerning nature and mechanisms whereby cannabinoids specifically alter susceptibility to infection by the ubiquitous opportunistic intracellular microbe *Legionella pneumophila* are then described in detail in the following chapter. Nicotine is now recognized as the addictive component of cigarette smoke and the next seven chapters review in detail studies concerning how nicotine affects the immune response, especially those aspects of immunity important in host resistance. It is widely recognized that cigarette smokers are more susceptible to upper respiratory infections by bacteria or viruses.

The next several chapters concern the effects of opiates on the immune system. Investigators from Temple University in Philadelphia describe studies concerning the effect of opiates on regulation of chemokine and chemokine receptor expression, known to be important in host resistance mechanisms, especially with emphasis on HIV infection. The next chapter by Dr. Roy and associates from Minnesota describes the effects of morphine on immune response mechanisms important in susceptibility to infections. Dr. Sharp and colleagues from Tennessee describe studies concerning neuropharmacological aspects of delta opioid receptors on murine splenic T cells and involvement of these receptors in immunity. Investigators from the University of Illinois then describe some of the effects of opiate derivatives on immunity, especially as related to mechanisms of resistance to infectious agents. The next several chapters discuss different aspects of the effects of alcohol on immunity, especially susceptibility to opportunistic bacteria and fungal infection. An experimental animal model is described concerning opportunistic infection by *Brucella* and ethanol. A general description of effects of alcohol on respiratory infections and the pulmonary system is then presented.

It is anticipated by the editors of this volume and the series in general, as well as the authors of individual chapters, that this book will be valuable for microbiologists, both basic and clinical, as well as immunologists, psychologists, and drug abuse investigators, including health care workers who care for and rehabilitate drug abusers. It is also anticipated that this book will also provide important information concerning the public health impact of drugs of abuse on infectious diseases. It is also hoped by the editors that the information presented will stimulate further interest and studies concerning the effects of drugs of abuse on infectious diseases. The editors thank Ms. Ilona M. Friedman for continued outstanding contributions as the coordinator for preparation of this volume and for valuable assistance in processing and editing manuscripts for publication.

Herman Friedman
Thomas W. Klein
Mauro Bendinelli

Contents

1

Effects of Cocaine and Morphine Withdrawal on the Immune Response

ALBERT H. AVILA, NORMA C. ALONZO,
and BARBARA M. BAYER

1. INTRODUCTION

The immunosuppression accompanying illicit drug use has been shown to contribute to a decreased resistance to a variety of pathogens; however, there is relatively little information on how long these effects persist following withdrawal from chronic drug exposure. To begin to address this question, Sprague–Dawley male rats were administered either cocaine (10 mg/kg, i.p., b.i.d.) for 7 days or morphine (escalating doses up to 40 mg/kg, s.c., b.i.d.) for a 10-day period. Control groups of animals received similar saline injections for equivalent time periods. Drug administration was abruptly discontinued and animals were sacrificed at 2, 24, 72, or 96 hr following the last dose. At these time points, proliferation responses of peripheral blood T lymphocytes stimulated by concanavalin A (Con A) and plasma levels of corticosterone were measured. Plasma corticosterone levels of cocaine- or morphine-treated animals were found to be significantly elevated 24 hr following drug cessation as compared to saline-treated animals. At this time, proliferation responses were significantly decreased and were further suppressed during cocaine and morphine withdrawal at 96 and 72 hr, respectively. These results suggest that abrupt cessation of cocaine or morphine administration leads to activation of stress-related

1

ALBERT H. AVILA, NORMA C. ALONZO, and BARBARA M. BAYER • Departments of Pharmacology and Neuroscience, Georgetown University Medical Center, 3900 Reservoir Road, N.W., Washington, DC 20007.

Infectious Diseases and Substance Abuse, edited by Herman Friedman *et al.*
Springer, New York, 2005.

pathways that may contribute to an increased susceptibility of infection during the initial withdrawal phase.

It is well known that cocaine and morphine abuse in general is a major health concern in our society. Studies have shown a high risk factor related to HIV seropositivity among cocaine users.[1,2] There appears to be an association between drug abuse populations and the development of AIDS, thus leading to the belief that the use of such drugs may serve as a cofactor in the pathogenesis of AIDS.[3,4] However, it is not clear if the immune alterations and susceptibility to AIDS is due to the lifestyle of the drug user (needles, nutrition, sexual practices) or to the effects of the drug itself.

Chronic cocaine and morphine exposure has been likened to the stress response due to their effect on the hypothalamic–pituitary–adrenal (HPA) axis resulting in elevation in plasma glucocorticoid levels.[5,6] As a result, many laboratories have investigated the potential interaction between HPA axis activation, stress and drug addiction, or relapse.[7–9] It is known that prolonged and permanent alterations within the central nervous system (CNS) occur following chronic cocaine or morphine administration,[6,10–12] as well as following withdrawal from either drug.[13–15] In addition to chronic exposure, abrupt withdrawal from chronic cocaine[16,17] or morphine[18] has also been shown to produce neuroendocrine alterations. Many of these effects have been thought to contribute to the immune deficiencies that accompany acute and chronic exposure to these drugs.[19–22] However, little is known of the potential impact that withdrawal from either morphine or cocaine has on the immune system. This is particularly surprising considering reports that cocaine abuse and dependence remains a major public health problem.[4] This is also surprising because many cocaine and drug abusers have a high potential for HIV exposure. If the immune system is compromised during the time of HIV exposure, the likelihood of higher viral titers and the susceptibility to contracting the disease is increased. Therefore, in this chapter, we begin to define the effects during cocaine or morphine exposure as well as during the early stages of withdrawal from chronic cocaine or morphine exposure on both the HPA axis and the immune system.

2. MATERIALS AND METHODS

2.1. Animals

Pathogen-free adult male Sprague–Dawley rats initially weighing 175–200 g upon receipt were obtained from Taconic Laboratories (Germantown, NY). Animals were group-housed, three per cage, with microisolator tops in a temperature $(23 \pm 1°C)$ and humidity-controlled vivarium under a 12-hr light/dark cycle (6 AM on, 6 PM off). Food and water were freely available (Purina rat chow, Ralston Purina Co., St. Louis, MO). All animals were allowed to acclimate for 1 week before use in an experiment or drug administration.

2.2. Drug Administration

Cocaine hydrochloride, purchased from Sigma Chemical (St. Louis, MO), and morphine sulfate, generously provided by the National Institute on Drug Abuse (Research Triangle Park, NC) were dissolved in (0.9%) sterile isotonic saline, which also served as the control treatment in these studies. The injection volume for both cocaine and morphine studies was 1 ml/kg and the route of administration was intraperitoneal (i.p.) for cocaine injections, and subcutaneous (s.c.) for morphine injections. For all cocaine injections, the rats received 10 mg/kg for 7 days (b.i.d.). For morphine injections, the animals were given escalating doses of morphine from 10 to 40 mg/kg for 9 days (b.i.d.), and were challenged with a 10 mg/kg injection of morphine on day 10. Animals were sacrificed 2 hr following the last cocaine injection or following respective withdrawal periods (24, 72, or 96 hr).

2.3. Mitogen-Induced Lymphocyte Proliferation

Rats were sacrificed by rapid decapitation, and trunk blood was collected in 50-ml polypropylene tubes containing heparin (0.1 ml) and immediately placed on ice. Whole blood was diluted 1:5 with cold RPMI-1640 media (Gibco BRL/Life Technologies, Grand Island, NY) containing 1% fetal calf serum and gentamicin (20 g/ml). Hundred liters of each blood suspension was plated into 96-well flat-bottom microtiter plates containing nine concentrations of the T-cell-specific mitogen Con A (100 L/well), incubated for 72 hr at 37°C with 8% CO_2 and pulsed with 0.5 Ci/well of [methyl-^{3}H]thymidine (6.7 Ci/mmol; New England Nuclear, Boston, MA) in a 20 L volume followed by additional 24 hr incubation. Cells were lysed by distilled water using a 96-well cell harvester (Brandel, Gaithersburg, MD), and labeled DNA was harvested onto glass fiber filters. Radioactivity was quantified via liquid scintillation spectrophotometry (Beta Plate; L.K.B. Pharmacia). Maximum lymphocyte proliferation responses (E_{max}) were determined from a nonlinear regression analysis of T-cell response to the mitogen Con A, and significant differences in E_{max} values were assessed using one-way ANOVA and Newman Keuls *post hoc* analysis.

2.4. Plasma Corticosterone Assay

Heparinized blood samples were collected at the time of sacrifice, placed on ice, and centrifuged to allow separation of plasma that was collected and stored at $-20°C$ until needed. Plasma corticosterone was measured using solid-phase double antibody ^{125}I radioimmunoassay kits purchased from ICN Biochemicals, Inc. (Costa Mesa, CA). Samples were assayed in duplicate, and corticosterone concentrations were expressed as nanograms per milliliter.

3. RESULTS

3.1. Immune and HPA Axis Effects from Acute Cocaine or Acute Morphine

As an initial assessment of cocaine's effects on the immune system, rats were injected with either acute cocaine (10 mg/kg, i.p.) or acute morphine (10 mg/kg, s.c.) and compared with saline control animals. All animals were sacrificed 2 hr following the injection, and blood was stimulated with increasing doses of the T-cell mitogen Con A. Maximum responses (E_{max}) were determined from a nonlinear regression analysis utilizing all concentrations of Con A, and significant differences in the E_{max} values were determined using the Student's *t*-test. T-lymphocyte proliferation did not differ between acute cocaine- and saline-treated animals (Fig. 1). In contrast to acute cocaine, acute morphine resulted in a significant suppression of blood lymphocyte proliferation ($p < 0.05$) (Fig. 2).

It is known that drugs of abuse can have stress-like effects on the HPA axis. To determine if there were any neuroendocrine effects, plasma corticosterone levels were measured in all animals. Animals were treated with either acute cocaine (10 mg/kg, i.p.) or saline and sacrificed 2 hr later. There were no significant differences in corticosterone levels at 2 hr between acute cocaine and saline control animals (Fig. 3). In contrast to cocaine, acute morphine (10 mg/kg, i.p.) led to a significant increase in plasma corticosterone levels 2 hr after a single morphine injection ($p < 0.05$) (Fig. 4).

3.2. Immune Effects Following Chronic Cocaine or Morphine

To determine whether chronic exposure to cocaine had effects on the immune system, animals were exposed to cocaine (10 mg/kg, i.p., b.i.p.) for 7 days. All animals were sacrificed 2 hr after the final injection. There was a

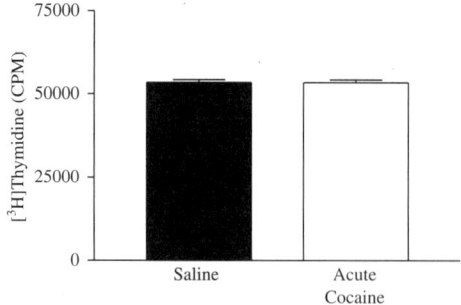

FIGURE 1. Effect of acute cocaine administration on blood lymphocyte proliferation. Animals (6 per group) were injected with cocaine (10 mg/kg, i.p.) or saline and sacrificed 2 hr following injection. Blood was collected into heparinized tubes, diluted 1:5 and lymphocyte proliferation stimulated by Con A. Data are expressed as E_{max} [³H]methyl-thymidine ± SEM. No significant difference in E_{max} values ($p > 0.05$, *t*-test).

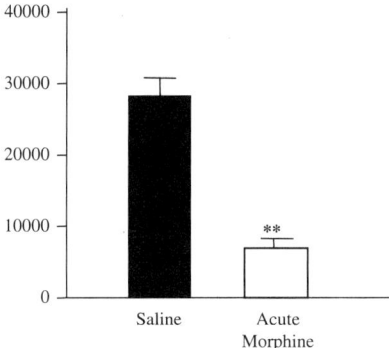

FIGURE 2. Effect of acute morphine administration on blood lymphocyte proliferation. Animals (8 per group) were injected with morphine (10 mg/kg, s.c.) or saline and sacrificed 2 hr following injection. Blood was collected into heparinized tubes, diluted 1:5 and lymphocyte proliferation stimulated by Con A. Data are expressed as E_{max} [^{3}H]methyl-thymidine $\pm$ SEM. Significant difference in E_{max} values ($p < 0.05$, t-test).

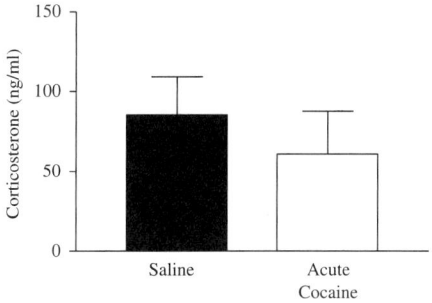

FIGURE 3. Effect of acute cocaine administration on plasma corticosterone. Animals (6 per group) were treated with either a single injection of cocaine (10 mg/kg, i.p.) or saline (1 ml/kg) and sacrificed 2 hr after injection via decapitation. Plasma corticosterone levels were determined as described in methods and expressed as mean (ng/ml) $\pm$ SEM. No significant alteration was detected ($P > 0.05$, Student's t-test).

significant decreased in T-cell proliferation in chronic cocaine animals compared to similarly treated saline controls (Fig. 5). Interestingly, following cessation of drug administration, this effect persisted for up to 96 hr following the last dose of cocaine ($p < 0.05$). Furthermore, animals that underwent 96 hr of withdrawal from cocaine were statistically more suppressed than those of the chronic cocaine group or the animals undergoing 24 hr of withdrawal.

Unlike chronic cocaine, chronic morphine treatment resulted in a tolerance to the suppressive effects of morphine on T-lymphocyte proliferation (Fig. 6). However, a significant suppression of lymphocyte responses developed within 24 hr after cessation of chronic morphine administration. The suppression of lymphocyte proliferation was significant for up to 72 hr of withdrawal from chronic morphine (Fig. 6).

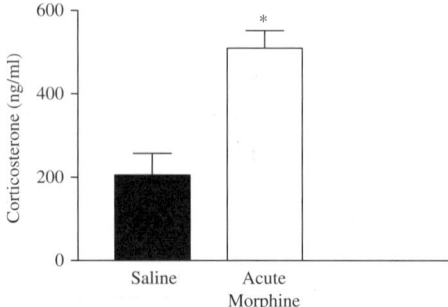

FIGURE 4. Effect of acute morphine administration on plasma corticosterone. Animals (6 per group) were treated with either a single injection of morphine (10 mg/kg, s.c.) or saline (1 ml/kg) and sacrificed 2 hr after injection via decapitation. Plasma corticosterone levels were determined as described in methods and expressed as mean (ng/ml) ± SEM. Significant alteration was detected ($P < 0.05$, Student's *t*-test).

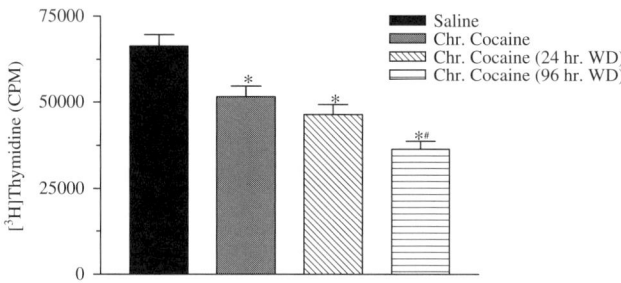

FIGURE 5. Effects of chronic cocaine treatment and abrupt withdrawal on lymphocyte proliferation. Animals were treated with either cocaine (10 mg/kg, i.p., b.i.d.) or saline for 7 days. Treatment groups were identical to those described in Fig. 1. Blood was collected into heparinized tubes, and lymphocyte proliferation responses to Con A were determined. Data are expressed as maximal responses ±SEM (CPM [^{3}H]methyl-thymidine per culture) of saline- and cocaine-treated animals ($n = 6$/treatment group). *Denotes significant difference ($p < 0.05$) compared to saline-treated control group (ANOVA, Newman Keuls *post hoc*). #Denotes significant difference ($p < 0.05$) compared to saline, chronic cocaine, and 24 hr WD groups (ANOVA, Newman Keuls).

3.3. Stress-Like Effects of Cocaine and Morphine Withdrawal

Animals treated with chronic cocaine for 7 days had corticosterone levels which were significantly elevated compared to saline controls. This was expected as it has been shown that tolerance does not develop to the HPA axis effects even after repeated cocaine injections.[25] Furthermore, a significant elevation of plasma corticosterone levels persisted in animals that went through the same chronic dosing paradigm, followed by 24 hr of withdrawal ($p < 0.05$). These findings are consistent with the report that cocaine withdrawal (12 hr) increases CRF release up to 400%.[17] Similarly, we found that corticosterone levels at 96 hr following withdrawal of cocaine returned to those of saline-treated animals

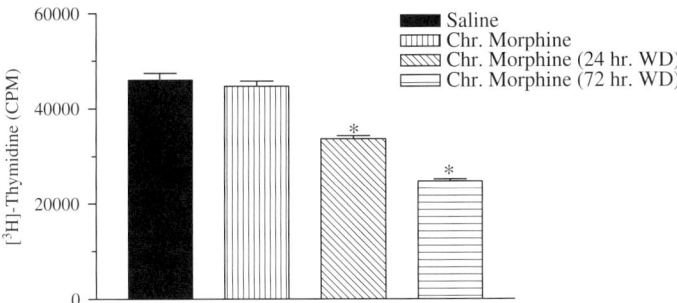

FIGURE 6. Effects of chronic morphine treatment and abrupt withdrawal on lymphocyte proliferation responses. Animals were treated with either saline or escalating doses of morphine (up to 40 mg/kg, s.c., b.i.d.) for 10 days. Treatment groups were the same as described in Fig. 3. Blood was collected into heparinized tubes, and lymphocyte proliferation to Con A was determined. Data are expressed as the maximum responses ± SEM (CPM [³H]methyl-thymidine per culture) of saline and morphine-treated animals ($n = 8$/treatment group). *Denotes significant difference ($p < 0.05$) compared to saline-treated control group (ANOVA, Newman Keuls).

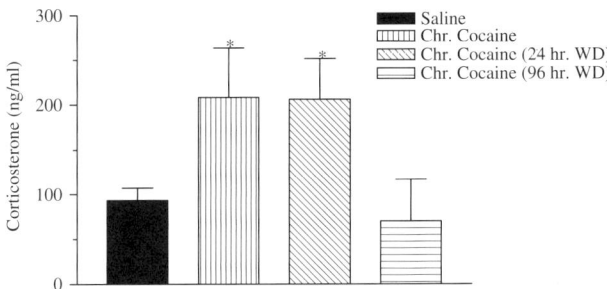

FIGURE 7. Changes in plasma corticosterone levels during the withdrawal period following cessation of chronic cocaine treatment. Animals were treated with either cocaine (10 mg/kg, i.p., b.i.d.) or saline for 7 days. The chronic cocaine treatment group was sacrificed 2 hr following the last cocaine injection (Chr. Cocaine). To induce withdrawal, cocaine administration was abruptly stopped and animals were sacrificed 24 hr (24 hr. WD) or 96 hr (96 hr. WD) later ($n = 6$/treatment group). Plasma samples were obtained and corticosterone levels were measured by RIA as described in Methods. *Denotes significant difference ($p < 0.05$) compared to saline-treated control group (ANOVA, Newman Keuls *post hoc*).

(Fig. 7). Therefore, these findings demonstrate that the initial cessation of drug administration results in an activation of the HPA axis, which is sustained for at least 24 hr.

In contrast to cocaine, animals chronically treated with morphine in escalating doses up to 40 mg/kg (s.c.) became tolerant to its effects on the HPA axis (Fig. 8). Other investigators have previously reported similar results.[24] However, when drug administration was abruptly discontinued, corticosterone levels were significantly elevated 24 hr later ($p < 0.05$). These findings are consistent with others who report that increases in glucocorticoids are observed upon sudden withdrawal from morphine administration.[18] Within 72 hr

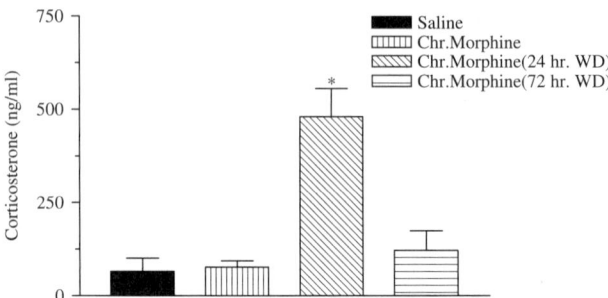

FIGURE 8. Effects of cessation of drug administration after chronic morphine treatment on plasma corticosterone levels. Animals were chronically treated with either saline or escalating doses of morphine (up to 40 mg/kg, s.c.) for up to 10 days. To initiate withdrawal, morphine administration was discontinued for either 24 or 72 hr prior to sacrifice. Plasma samples were obtained and corticosterone levels were measured by RIA as described in *Methods*. *Denotes significant difference ($p < 0.05$) compared to the saline-treated control group (ANOVA, Newman Keuls *post hoc*).

following withdrawal of morphine, corticosterone values returned to basal levels compared to saline controls.

4. DISCUSSION

The studies described in this chapter examine and provide significant information concerning the effects of cocaine and morphine withdrawal on the immune response of rats. Namely, withdrawal from chronic cocaine or chronic morphine lead to activation of the HPA axis resulting in increased glucocorticoid levels, which was accompanied by a suppression of T-lymphocyte proliferation. Suppression of the immune response was observed following chronic cocaine, but tolerance developed to chronic morphine treatment. Whereas acute cocaine did not lead to immunosuppression, acute morphine treatment suppressed the immune response.

It has been reported that acute cocaine dose-dependently increases corticosterone levels in rats,[25] as well as stimulates the release of ACTH and cortisol in humans.[26,27] It was found that a single acute cocaine injection rapidly and transiently increases corticosterone levels within the first 30 min postinjection, and returns to baseline values by 2 hr.[28] In this study, animals were sacrificed at 2 hr following acute cocaine; perhaps this is why an increase in plasma corticosterone levels following acute treatment was not detected. In addition, acute morphine has also been labeled a "pharmacological" stress due to its ability to induce behavioral, neural, and endocrine alterations.[6,29,30] There was an increase in plasma corticosterone levels following acute morphine treatment in animals assayed 2 hr following drug exposure.

Consistent with our finding that 7 days of cocaine exposure significantly increased basal corticosterone levels 2 hr after the last cocaine injection, others have reported increases in basal levels of corticosterone after 3 weeks of cocaine administration.[23] Increases in plasma corticosterone have also been reported in rats trained to self-administer cocaine.[31] These physiological "anxiety"-like effects, namely the reported increases in ACTH and corticosterone due to cocaine exposure, have been blocked in rats pretreated with the CRF receptor antagonist, alpha-helical CRF[9–41] or with an anti-CRF antibody.[25,32] Withdrawal from cocaine or morphine resulted in a significant and prolonged elevation in steroid levels that were sustained for at least 1 day after the cessation of chronic treatment with either drug, and were back near baseline values by 2 days of withdrawal. These results are consistent with two other studies, which have reported prolonged increases (12–24 hr) in corticosterone levels after cessation of cocaine.[24,33] More recently, it has been reported that cocaine withdrawal increases CRF release up to 400% above baseline levels in the amygdala between 11 and 12 hr post-cocaine,[17] which supports our finding of a prolonged glucocorticoid response.

Although some reports have demonstrated that cocaine and morphine administrations produce alterations in a variety of immunological parameters,[19–22,29,34] there is comparatively little information on whether the immune alterations persist during chronic drug administration and after cessation of either cocaine or opioid administration. These data illustrate several similarities in the effects following withdrawal from cocaine and morphine. The cessation of both drugs of abuse resulted in elevation of plasma corticosteroid levels for the first 24 hr of withdrawal. This effect was accompanied by a significant suppression of the immune system to both cocaine and morphine. With both drugs, immunosuppression persisted for 3–4 days even though corticosterone values had returned to basal levels. Although this report demonstrates that there is a stress-like effect from withdrawal of cocaine and morphine seen by HPA-axis activation, the precise mechanisms involved in producing the prolonged effects on the immune system remain to be determined.

There are several possible explanations which may contribute to the prolonged suppression of the T-lymphocyte immune response following withdrawal from either chronic cocaine or chronic morphine. One reason may be that the intermittent increases and decreases of corticosterone throughout the drug administration period could lead to immune system vulnerability and ultrasensitivity during the subsequent drug withdrawal phase. Another explanation may be due to the sustained increase in corticosterone during the withdrawal period.[23] For over two decades, corticosteroids have been known to decrease immune cell function.[35] It is possible that both the repeated elevation during chronic dosing as well as the prolonged increase in corticosterone during withdrawal may contribute to immunosuppression. Additionally, corticosterone may cause an initial priming effect during chronic dosing, followed by a prolonged increase in corticosterone levels from the stress of withdrawal, resulting in sustained suppression in T-cell proliferation.

In conclusion, these data indicate that during the withdrawal period, there are increases in steroid levels accompanied by profound and prolonged suppressive effects on the immune system, which may lead to an increased susceptibility to infection. As a result, it will be important to further characterize the duration of these effects and to more completely assess the immune vulnerability during the withdrawal period. Ultimately, these studies may lead to a possible model for testing potential pharmacotherapies used in relapse and drug abuse patients.

REFERENCES

1. Anthony, J. C., Vlahov, D., Nelson, K. E., Cohn, S., Astemborski, J., and Solomon, L. (1991). New evidence on intravenous cocaine use and the risk of infection with human immunodeficiency virus type 1. *Am. J. Epidemiol.* **134**(10):1175–1189.
2. Chaisson, R. E., Bacchetti, P., Osmond, D., Brodie, B., Sande, M. A., and Moss, A. R. (1989). Cocaine use and HIV infection in intravenous drug users in San Francisco. *JAMA* **261**(4):561–565.
3. Donahoe, R. M. and Falek, A. (1988). Neuroimmunomodulation by opiates and other drugs of abuse: Relationship to HIV infection and AIDS. *Adv. Biochem. Psychopharmacol.* **44**:145–158.
4. Donahoe, R. M. (1990). Drug abuse and AIDS: Causes for the connection. *NIDA Res. Monogr.* **96**:181–191.
5. Sarnyai, Z., Mello, N. K., Mendelson, J. H., Eros-Sarnyai, M., and Mercer, G. (1996). Effects of cocaine on pulsatile activity of hypothalamic–pituitary–adrenal axis in male rhesus monkeys: Neuroendocrine and behavioral correlates. *J. Pharmacol. Exp. Ther.* **277**(1):225–234.
6. Houshyar, H., Galignîana, M. D., Prat, W.B., and Woods, J.H. (2001). Differential responsivity of the hypothalamic pituitary-adrenal axis to glucocorticoid negative-feedback and corticotropin releasing hormone in rats undergoing morphine withdrawal: Possible mechanisms involved in facilitated and attenuated stress responses. *J. Neuroendocrinol.* **13**:875–886.
7. Piazza, P. V. and Le Moal, M. (1998). The role of stress in drug self-administration. *Trends Pharmacol. Sci.* **19**(2):67–74.
8. Koob, G. F. (1999). Stress, corticotropin-releasing factor, and drug addiction. *Ann. NY Acad. Sci.* **897**:27–45.
9. Goeders, N. E. (2002). Stress and cocaine addiction. *J. Pharmacol. Exp. Ther.* **301**(3):785–789.
10. Self, D. W. and Nestler, E. J. (1995). Molecular mechanisms of drug reinforcement and addiction. *Annu. Rev. Neurosci.* **18**:463–495.
11. Berke, J. D. and Hyman, S. E. (2000). Addiction, dopamine, and the molecular mechanisms of memory. *Neuron* **25**(3):515–532.
12. Gonzalez, S., Fernandez-Ruiz, J., Sparpaglione, V., Parolaro, D., and Ramos, J.A. (2002). Chronic exposure to morphine, cocaine or ethanol in rats produced different effects in brain cannabinoid CB(1) receptor binding and mRNA levels. *Drug Alcohol Depend.* **66**:77–84.
13. Kuhar, M. J. and Pilotte, N. S. (1996). Neurochemical changes in cocaine withdrawal. *Trends Pharmacol. Sci.* **17**(7):260–264.
14. Fuentealba, J. A., Forray, M. I., and Gysling, K. (2000). Chronic morphine treatment and withdrawal increase extracellular levels of norepinephrine in the rat bed nucleus of the stria terminalis. *J. Neurochem.* **75**:741–748.
15. Devoto, P., Flore, G., Pira, L., and Gessa, L. (2002). Co-release of noradrenaline and dopamine in the prefrontal cortex after acute morphine and during morphine withdrawal. *Psychopharmacoly (Berl)* **160**:220–224.
16. Avila, A. H., Morgan, C. A., and Bayer, B. M., (2003). Stress-induced suppression of the immune system after withdrawal from chronic cocaine. *J. Pharmacol. Exp. Ther.* **305**: 290–297.
17. Richter, R. M. and Weiss, F. (1999). In vivo CRF release in rat amygdala is increased during cocaine withdrawal in self-administering rats. *Synapse* **32**(4):254–261.

18. Kishioka, S., Nishida, S., Fukunaga, Y., and Yamamoto, H. (1994). Quantitative properties of plasma corticosterone elevation induced by naloxone-precipitated withdrawal in morphine-dependent rats. *Jpn J. Pharmacol.* **66**:257–263.

19. Van Dyke, C., Stesin, A., Jones, R., Chuntharapai, A., and Seaman, W. (1986). Cocaine increases natural killer cell activity. *J. Clin. Invest.* **77**(4):1387–1390.

20. West, J.P., Lysle, D. T., and Dykstra, L.A. (1997). Tolerance development to morphine-induced alterations of immune status. *Drug Alcohol Depend.* **46**:147–157.

21. Pellegrino, T. and Bayer, B. M. (1998). In vivo effects of cocaine on immune cell function. *J. Neuroimmunol.* **83**:139–147.

22. Mellon, R. D. and Bayer, B. M. (1999). The effects of morphine, nicotine and epibatidine on lymphocyte activity and hypothalamic–pituitary–adrenal axis responses. *J. Pharmacol. Exp. Ther.* **288**:635–642.

23. Sarnyai, Z. (1998). Neurobiology of stress and cocaine addiction. Studies on corticotropin-releasing factor in rats, monkeys, and humans. *Ann. NY Acad. Sci.* **851**:371–387.

24. Borowsky, B. and Kuhn, C. M. (1991). Chronic cocaine administration sensitizes behavioral but not neuroendocrine responses. *Brain Res.* **543**(2):301–306.

25. Sarnyai, Z., Hohn, J., Szabo, G., and Penke, B. (1992). Critical role of endogenous corticotropin-releasing factor (CRF) in the mediation of the behavioral action of cocaine in rats. *Life Sci.* **51**(26):2019–2024.

26. Mendelson, J. H., Teoh, S. K., Mello, N. K., Ellingboe, J., and Rhoades, E. (1992). Acute effects of cocaine on plasma adrenocorticotropic hormone, luteinizing hormone and prolactin levels in cocaine-dependent men. *J. Pharmacol. Exp. Ther.* **263**(2):505–509.

27. Baumann, M. H., Gendron, T. M., Becketts, K. M., Henningfield, J. E., Gorelick, D. A., and Rothman, R. B. (1995). Effects of intravenous cocaine on plasma cortisol and prolactin in human cocaine abusers. *Biol. Psychiatry* **38**(11):751–755.

28. de Kloet, E. R. (1992). Corticosteroids, stress, and aging. *Ann. NY Acad. Sci.* **663**: 357–371.

29. Buckingham, J. C. and Cooper, T. A. (1984). Differences in hypothalamo–pituitary–adrenocortical activity in the rat after acute and prolonged treatment with morphine. *Neuroendocrinology* **38**(5):411–417.

30. Martinez-Pinero, M. G., Milanes, M. V., Alcaraz, C., and Vargas, M. L. (1994). Catecholaminergic mediation of morphine-induced activation of pituitary-adrenocortical axis in the rat: Implication of alpha- and beta-adrenoceptors. *Brain Res.* **668**(1–2):122–128.

31. Goeders, N. E. and Clampitt, D. M. (2002). Potential role for the hypothalamo–pituitary–adrenal axis in the conditioned reinforcer-induced reinstatement of extinguished cocaine seeking in rats. *Psychopharmacology (Berl)* **161**(3):222–232.

32. Rivier, C. and Vale, W. (1987). Cocaine stimulates adrenocorticotropin (ACTH) secretion through a corticotropin-releasing factor (CRF)-mediated mechanism. *Brain Res.* **422**(2):403–406.

33. Levy, A. D., Rittenhouse, P. A., Li, Q., Yracheta, J., Kunimoto, K., and Van de Kar, L. D. (1994). Influence of repeated cocaine exposure on the endocrine and behavioral responses to stress in rats. *Psychopharmacology (Berl)* **113**(3–4):547–554.

34. Klein, T. W., Newton, C., and Friedman, H. (1991). Cocaine effects on cultured lymphocytes. *Adv. Exp. Med. Biol.* **288**:151–158.

35. Parrillo, J. E. and Fauci, A. S. (1979). Mechanisms of glucocorticoid action on immune processes. *Annu. Rev. Pharmacol. Toxicol.* **19**:179–201.

2

Biological Links between Cannabinoids and HIV Infection

GAYLE C. BALDWIN and MICHAEL D. ROTH

1. INTRODUCTION

Infection with human immunodeficiency virus (HIV) and progression of the acquired immunodeficiency syndrome (AIDS) can be modulated by a variety of cofactors including genetic susceptibility, nutritional factors, and the presence of concurrent infections.[1–5] Marijuana, cocaine, alcohol, and other substances of abuse may also be risk factors for HIV,[6–12] both in terms of their influence on risky social behavior[13,14] as well as their potential to alter host immunity and viral replication.[15–20] Marijuana use is prevalent among homosexual and bisexual men at risk for acquiring HIV, and occurs at an even higher frequency in those who develop HIV infection.[21] In addition to recreational and social use, marijuana is also used as a medicinal agent for the nausea, pain, and wasting states that occur in AIDS.[22,23] This frequent use of marijuana by individuals with HIV and/or AIDS may be clinically important. Marijuana smoking has been reported as a risk factor for the development of bacterial pneumonia, opportunistic infections, and Kaposi's sarcoma in HIV-positive individuals,[9,24] as well as for a more rapid progression from HIV infection to AIDS.[11] At the same time, cannabinoids have been postulated to play a neuroprotective role by suppressing

GAYLE C. BALDWIN • Division of Hematology/Oncology, Department of Medicine, David Geffen School of Medicine at UCLA, Los Angeles, CA 90095.

MICHAEL D. ROTH • Division of Pulmonary and Critical Care, Department of Medicine, David Geffen School of Medicine at UCLA, Los Angeles, CA 90095-1690.

Infectious Diseases and Substance Abuse, edited by Herman Friedman *et al.* Springer, New York, 2005.

the production of inflammatory mediators and nitric oxide (NO), potentially reducing HIV-related neurotoxicity and the development of AIDS-related dementia.[25–27] This chapter will focus on biological links between marijuana use, cannabinoid immunobiology, and the pathogenesis of HIV infection. We will review information regarding the expression of cannabinoid receptors on brain and immune cells, the capacity for Δ^9-tetrahydrocannabinol (THC) to suppress T-cell immunity and alter cytokine production, the impact of cannabinoids on the production of NO, and the potential link of these effects to the pathogenesis of HIV infection and AIDS. We will also introduce a mouse model in which THC can be directly evaluated for its effect on the interaction between human immune cells, HIV infection, and viral replication *in vivo*.

2. MARIJUANA SMOKE, CANNABINOIDS, AND CANNABINOID RECEPTORS

The tar produced from marijuana smoke contains a high concentration of cannabinoids.[28–30] Classical cannabinoids are a structurally related group of C_{21}-hydrocarbons obtained exclusively from the plant *Cannabis sativa*.[31] While 61 different natural cannabinoids have been described, THC is the predominant form in marijuana and is primarily responsible for its biologic activity.[28] Synthetic THC, called dronabinol, is also the active component in the prescription drug Marinol®. Dronabinol is approved by the Food and Drug Administration for the treatment of anorexia and weight loss in patients with AIDS and for refractory nausea and vomiting associated with cancer chemotherapy.[22,23] In addition to these exogenous sources, several endogenous compounds with cannabinoid-like activity have been identified within the brain and peripheral tissues.[32–35] Although structurally distinct from THC, these endocannabinoids bind cannabinoid receptors and induce similar biological effects. A list of the cannabinoids relevant to the discussion of HIV infection is presented in Table I.

Two structurally related receptors that bind THC with similar affinity have been described. The gene encoding for cannabinoid receptor 1 (CB1) was cloned in 1990,[36,37] and this receptor is highly expressed in brain tissue and to a lesser extent in the adrenal glands, reproductive organs, and on immune cells.[38,39] Animal models employing selective CB1 antagonists or CB1 knockout mice have demonstrated that CB1 receptors mediate most of the psychoactive, behavioral, and physiologic effects commonly associated with marijuana use.[40] The role of CB1 in these responses has also been confirmed in humans. Research subjects pretreated with a CB1-selective receptor antagonist (SR141716) do not develop the psychological high or tachycardia normally produced by marijuana use.[41] A second cannabinoid receptor, CB2, was cloned from a human promyelocytic cell line.[42] While not normally expressed in the brain, CB2 is predominantly expressed in spleen and in cells of hematopoietic origin.[38] Using semi-quantitative RT-PCR, Nong and coworkers[43] recently confirmed that mRNA encoding for both CB1 and CB2 are present in normal human peripheral blood mononuclear cells, that expression of CB2 is uniformly 3-fold higher than

TABLE I
Cannabinoids and Cannabinoid Receptors Relevant to
AIDS and HIV Infection

		Affinity for cannabinoid receptors	
Exposure	Primary cannabinoids	Brain (CB1)	Immune cells (CB2)
Marijuana smoke	Δ^9-tetrahydrocannabinol (THC)	Yes	Yes
Marinol®	Synthetic THC (dronabinol)	Yes	Yes
Endogenous ligands	N-arachidonoylglycerol (anandamide)	Yes	Weak
	2-arachidonoylglycerol (2-AG)	Yes	Yes
	2-Arachidonyl ether (nolander ether)	Yes	Minimal

expression of CB1, and that mRNA for both receptors is upregulated in samples collected from habitual marijuana users. B cells express the highest levels of CB2 mRNA, with somewhat less expression found in natural killer cells and monocytes, and lower levels of expression present in CD8+ and CD4+ T cells.[39,44] Both CB1 and CB2 are seven transmembrane G-protein-coupled receptors.[45] Their activation blocks the forskolin-induced accumulation of intracellular cyclic adenosine 3′,5′-monophosphate (cAMP) in immune cells and is associated with a variety of downstream signaling events including inhibition of calcium flux and protein kinase A (PKA), downregulation of activator protein-1 (AP-1) and the nuclear factor of activated T cells (NF-AT), alterations in MAP kinase signaling, and changes in binding of cAMP response elements (CRE) and the nuclear factor for immunoglobulin k chain (NF-κB).[46,47] The role of these receptors in human biology is still an area of intense investigation. Both neurons and immune cells produce endogenous CB1 and CB2 ligands, such as N-arachidonoylglycerol (anandamide), 2-arachidonoylglycerol (2-AG), and 2-arachidonyl ether (nolander ether), suggesting an important role for the cannabinoid ligand-receptor pathway in regulating normal function within both the central nervous system and the immune system.[20,35,48–51] It is the expression of CB1 and CB2 on immune cells and within the brain that likely mediates the interaction between cannabinoids, host responses, and HIV infection.

3. CANNABINOIDS REGULATE CYTOKINE PRODUCTION, T-CELL ACTIVATION, AND HOST IMMUNE RESPONSES IN A MANNER THAT MAY PROMOTE HIV INFECTION AND THE PROGRESSION OF AIDS

HIV infection produces wide-ranging effects on the human immune system, hallmarks of which include a sustained release of acute inflammatory mediators such as tumor necrosis factor-alpha (TNF-α) and IL-6,[52,53] increased levels of immunosuppressive cytokines including transforming growth factor-beta (TGF-β) and IL-10,[54–56] polyclonal activation of B cells,[54,57] a relative deficiency in the production of interferon-gamma (IFN-γ) and IL-12,[58,59] and the destruction/functional impairment of T cells,[60,61] antigen-presenting cells,[62,63] and antigen-specific immunity.[64] The biological consequences of these changes include the permissive expansion and replication of HIV, inflammatory injury to bystander tissues and organs such as the central nervous system, and a profound cellular immunodeficiency that allows the development of both opportunistic infections and cancer. Theoretically, biological agents that promote a similar pattern of immune dysfunction will enhance HIV infection and the progression of AIDS. The immunologic effects of cannabinoids have been reviewed before[16,20,46,50,65–67] and are briefly summarized in the following section.

Marijuana and THC were first reported to modulate immune function in the 1970s when abnormal T-cell responses were observed in THC-treated animals[68] and in peripheral blood mononuclear cells (PBMC) collected from chronic marijuana smokers.[69] T-cell proliferation was reduced in mixed leukocyte and mitogen-stimulation assays by 40–45% when a group of 51 marijuana smokers were compared to nonsmoking controls.[69] However, the link between THC and immune regulation was not widely accepted until the discovery that leukocytes express cannabinoid receptors.[38,39,42] The preferential expression of cannabinoid receptors on immune cells, the ability to regulate leukocyte function with receptor-specific ligands, and the capacity to block receptor activation with selective antagonists, have all provided important insight into the role of cannabinoids as immune regulators.[50,66]

A variety of mouse models have been used to demonstrate the suppressive effects of THC on host defenses; effects which allow the propagation of viral infections,[65] opportunistic infections,[15,70] and cancer.[71] Newton and associates[70] treated mice with a single 4 mg/kg dose of THC prior to inoculation with a sublethal dose of *Legionella pneumophila*. When subsequently challenged with a lethal bacterial load, control mice demonstrated antigen-specific immunity and eradicated the infection. In contrast, the majority of animals pretreated with THC during the immunization phase died following rechallenge and their T cells failed to proliferate in response to *L. pneumophila* antigen *in vitro*. T cells, and the cytokines that they produce, serve as critical regulators of cell-mediated immunity. T cells producing IL-2 and IFN-γ (T helper 1 subtype, Th1) stimulate macrophage and T-cell effecter function and promote cell-mediated immunity.[72] In contrast, T cells producing primarily IL-4 and IL-10 (T helper 2 subtype, Th2), suppress cell-mediated immunity and promote humoral and allergic responses. Consistent with a switch from a Th1 to a Th2 response, THC was found to downregulate the production of anti-legionella antibody of the IgG$_{2a}$ subclass and increase antibody of the IgG$_1$ subclass. *In vitro*, control splenocytes activated with immobilized anti-CD3 antibody secreted primarily IFN-γ with little IL-4. In contrast, splenocytes activated in the presence of THC produced the opposite profile with less IFN-γ and more IL-4.[70] The capacity for THC to block immunity against *L. pneumophila*, promote an immunoglobulin isotype switch from IgG_{2a} to IgG_1, and alter the balance of memory T cells producing Th1 and Th2 cytokines, provided the first evidence that cannabinoids act as Th1:Th2 modulators, promoting a relative predominance of Th2 cytokines. In follow-up experiments,[15] pretreatment with THC prior to infection with *L. pneumophila* was found to result in lower serum concentrations of IL-12 and IFN-γ, and primed splenocytes to secrete higher levels of IL-4 within hours after infection. Administration of either a selective CB1 or CB2 receptor antagonist (SR141716A or SR144528, respectively) blocked the effects of THC, confirming the role of cannabinoid receptor signaling in mediating the immunologic consequences of THC.

Cell-mediated immunity and Th1/Th2 cytokine balance also play a central role in limiting tumor growth.[73,74] Zhu and associates[71] pretreated mice with daily intraperitoneal injections of THC (5 mg/kg) for 4 days each week and then challenged them with subcutaneous tumor implants. As one might hypothesize from the infection models, mice receiving THC experienced a more rapid rate of tumor growth. By the end of 5–6 weeks, tumors in THC-treated animals averaged 3–4 times the size as did tumors growing in control animals.[71] Since there was no direct effect of THC on the proliferation of tumors *in vitro,* these studies suggested that THC might enhance tumor growth by disrupting immune function *in vivo.* Consistent with this, splenocytes produced less IFN-γ when harvested from tumor-bearing mice that had been treated with THC in comparison to tumor-bearing controls. In addition, T cells recovered from THC-treated animals produced higher concentrations of IL-10 and TGF-β and proliferated poorly when stimulated by allogeneic dendritic cells. There were also specific defects in antigen-presenting activity when dendritic cells harvested from THC-treated mice were compared to control dendritic cells for their capacity to activate T cells. A central role for TGF-β and IL-10 in mediating the adverse effects of THC was demonstrated by the administration of neutralizing antibodies specific for these two cytokines, either of which completely prevented the effects of THC on tumor growth.[71] The biologic effects of THC in this model were also blocked by the administration of a CB2 receptor antagonist.

In order to evaluate whether human immune responses are effected in the same way by THC, T cells have been collected from healthy volunteers and examined *in vitro.*[75,76] Whether added to mixed leukocyte reaction (MLR) assays or to T cells stimulated by immobilized anti-CD3 and anti-CD28 monoclonal antibodies, THC suppressed T-cell proliferation, downregulated the expression and release of Th1 cytokines, increased the expression of Th2 cytokines, and altered normal Th1/Th2 balance in a dose-dependent manner.[76] THC inhibited the proliferation of antigen-specific T-cell clones, with 5 µg/ml inhibiting activation by an average of 53% compared to control T cells exposed to diluent alone. IFN-γ concentrations were reduced on average by 50%, while IL-4 levels were increased on average to 110%, resulting in a shift in Th1/Th2 cytokine balance. These results were strikingly similar to the downregulation of antigen-specific Th1 cells and the upregulation of antigen-specific Th2 cells observed in the intact animal models.[15,70,71] The impact of THC on Th subsets was also examined at the level of mRNA expression using a ribonuclease protection assay to simultaneously assay for both Th1 (IL-2, IFN-γ) and Th2 (IL-4, IL-5) cytokines.[76] Consistent with the results obtained by ELISA, mRNA transcripts encoding for IFN-γ and IL-2 were reduced by 20–50% in cells treated with THC, and mRNA encoding for IL-4 and IL-5 were increased up to 11-fold. Pretreatment with SR144528, a CB2-selective antagonist, prevented the majority of the THC-mediated effects, while there was little response to AM251, a selective CB1 antagonist. This work suggests a strong correlation between murine models and human studies, with THC acting via cannabinoid receptors to suppress antigen-specific T-cell activation and skew responding T cells away from a Th1 response and toward a Th2 profile.

THC has also been shown to upregulate the production of TGF-β when human T cells are activated by signaling through the T-cell receptor.[75] TGF-β inhibits T-cell proliferation, suppresses production of IL-2 and IFN-γ, and antagonizes the activation of both lymphocytes and monocytes. As little as 50 ng/ml of THC increased the production of TGF-β by 2–3-fold and 5 μg/ml of THC increased the release of TGF-β protein by 5-fold. Selective CB1 or CB2 receptor antagonists were used to confirm that signaling was mediated via the CB2 receptor.[75]

In addition to the coordinated regulation of Th1/Th2 cytokines and the suppression of T-cell-based immunity, cannabinoids have also been shown to modulate the production of inflammatory mediators, including IL-1 and TNF-α,[26,77–79] B-cell activation,[80,81] and the induction of apoptosis in T cells.[82–84] However, the effects of cannabinoids on these functions appear to be variable, depending upon the model employed and the concentrations of cannabinoid studied. In early studies, direct injection of THC into mice in conjunction with inflammatory or infectious stimuli produced high levels of the acute-phase mediators IL-1, IL-6, and TNF-α.[78,84] In this situation, administration of THC was associated with increased mortality. However, when studied with isolated cells *in vitro*, THC produced a dose-dependent decrease in the production of IL-1 and TNF-α, an effect associated with protection against neurotoxicity.[26,77] At low concentrations, THC and other CB2-active cannabinoids increase the proliferation of activated B cells, but at higher concentrations, they suppressed B-cell activation.[80] When examined in detail in the mouse and with murine splenocytes *in vitro*, the administration of THC suppressed humoral responses to T-cell-dependent antigens (sheep red blood cells), but had no effect on B-cell responses to T-cell independent responses (DNP antigen) or to polyclonal activation by lipopolysaccharide (LPS).[81] Cannabinoids have also been reported to directly stimulate apoptosis in leukocytes, including activated T cells and a variety of lymphoid leukemia cells.[82–84] Zhu and associates[84] were able to directly relate THC to the induction of DNA fragmentation and strand breaks in LPS or Con-A activated splenocytes, downregulation of Bcl-2, and activation of Caspase-1. Apoptosis was blocked by a caspase-1 inhibitor. McKallip and coworkers[82] observed apoptotic cell death in several human and murine lymphoblastoid or leukemic cell lines, as well as in fresh leukemic cells, when exposed to THC either *in vitro* or *in vivo*. In all cases, these effects were blocked by a CB2 receptor antagonist.

Together, these human and animal models provide important insight into the impact of THC on host immunity and the potential interaction between cannabinoids and HIV infection (Table II). The integrated downregulation of cell-mediated immunity by THC is likely to synergize with the immunosuppressive effects mediated by HIV. In epidemiologic studies, marijuana use is associated with an increased risk for opportunistic bacterial and parasitic pneumonia in HIV patients,[24] an increased incidence of HIV infection and more rapid progression from HIV infection to AIDS.[9,11] Deficiencies in the production of IFN-γ and IL-12, which are uniformly observed in response to THC, also play a central role in AIDS-related immunodeficiency and the incapacity for HIV-infected patients to respond to infectious pathogens or vaccines.[59] Similarly, TGF-β and IL-10 are induced in response to both THC and HIV viral proteins such as Tat and gp120.[54,85–87] Increased levels of TGF-β are implicated in the suppression of IFN-γ and IL-12, may contribute to T-cell apoptosis in combination with HIV infection,[88] promote fibrosis and HIV-nephropathy,[89] and amplify HIV replication.[18,90,91] The capacity for THC to induce lymphocyte apoptosis[82,92] might synergize with the apoptotic effects mediated by TGF-β, viral proteins, and other factors during HIV infection, thereby promoting the loss of helper T cells. The role of IL-10 in HIV is more controversial. IL-10 may directly suppress host immunity,[55] promote B-cell activation and B-cell

TABLE II
Links between the Biology of Cannabinoids and the Pathogenesis of HIV

Biologic effects of cannabinoids	Consequences for patients with HIV	
	Potential positive effects	Potential negative effects
Downregulation of IL-2, IFN-γ, and IL-12 (Th1 cytokines)		Inhibit function of monocytes, T-, NK-, and dendritic cells Depress cellular immunity
Upregulation of IL-4, IL-5, and IL-10 (Th2 cytokines)	Limit proinflammatory cytokines involved in tissue injury (IL-1, IL-6, TNF-α) Suppress HIV replication	Suppress cellular immunity Increase allergy/atopy B-cell activation and proliferation Upregulate HIV co-receptors
Production of TGF-β		Suppression of IFN-γ/IL-12 Promote T-cell apoptosis Promote fibrosis/nephropathy Enhance replication of HIV
Suppress antigen presenting cells and cellular immunity		Promote opportunistic infections Promote tumor growth
B-cell activation		Enhance polyclonal gammopathy Promote B-cell malignancies
T-cell apoptosis		Synergize with HIV to destroy activated T cells
Inhibit production of nitric oxide	Decrease HIV replication Protect against neuronal injury and AIDS dementia	Reduce antimicrobial defenses Suppress antiviral responses Increase monocyte/macrophage viral reservoir

lymphomas,[54] and increase HIV replication by increasing expression of HIV co-receptors on monocytes and T cells.[93,94] A potential interaction between THC-induced upregulation of IL-10, its capacity to enhance B-cell proliferation, and the pathogenesis of AIDS-related B-cell lymphoma should be considered. Alternatively, IL-10 may limit the induction of proinflammatory cytokines such as IL-6 and TNF-α, suppress T-cell activation, and thereby limit HIV progression and HIV-associated tissue injury.[54,95] More information is required about the impact of marijuana and THC on cytokine production and secondary consequences in HIV patients before any conclusions can really be drawn in this respect. Finally, one needs to consider the role of THC and endogenous cannabinoids as potentially beneficial agents in suppressing acute inflammatory cytokine release. In this one respect, HIV and THC appear to differ in their impact on immune function. HIV is associated with overproduction of IL-1 and TNF-α that might be associated with AIDS-related wasting, apoptosis, and neurological injury.[27,54,96] Several models suggest that both endogenous cannabinoids and exogenous administration of THC may reduce TNF-α and protect against these adverse effects of HIV on the central nervous system.[25–27,97] Again, THC can regulate a variety of cytokines and effecter cells that are directly relevant to the pathogenesis of HIV and more information is required in order to understand the complex interaction between this drug, HIV infection, and the progression of AIDS.

4. MARIJUANA AND THC SUPPRESS THE INDUCTION OF NITRIC OXIDE WITH THE POTENTIAL FOR BOTH POSITIVE AND NEGATIVE EFFECTS ON THE HOST RESPONSE TO HIV

NO is a signaling molecule implicated in a diverse repertoire of regulatory functions ranging from neurotransmission to vasodilation and blood pressure control.[98–100] NO also plays a key role as an immune effecter and signaling molecule and represents an important component of the host immune response against bacteria, protozoa, tumor cells, and viruses.[101–104] Although classically viewed as a proinflammatory mediator that protects against infectious agents, NO has a particularly complex role in defense against viruses, especially HIV.[27,105–108] Studies have shown that production of NO is elevated in HIV-infected individuals and in acute simian immunodeficiency (SIV) models.[97,109–112] Several factors may be involved in stimulating NO overproduction during HIV infection including direct stimulation by viral proteins such as gp120 and Tat, the chronic elevation of proinflammatory cytokines such as IL-1 and TNF-α, and stimulation by opportunistic infections.[27] While NO, acting either directly and/or indirectly, may play a protective role by destroying HIV-infected cells or blocking HIV replication through inhibition of viral enzymes, reverse transcriptases, proteases, or cellular transcription factors,[113–118] it may also contribute to adverse effects in HIV-infected patients. Both *in vitro* studies and human studies suggest that overproduction of NO can contribute to HIV replication, increase HIV-associated immune suppression, and mediate cytotoxic effects on neural tissue.[27,119–122] Jimenez and coworkers[97] found that addition of NO donors to activated peripheral blood mononuclear cultures significantly enhanced the replication of HIV, and that treatment with inhibitors of inducible NO synthase (iNOS) suppressed viral replication. In that study, NO specifically enhanced transcription from long terminal repeat elements that are active during stimulation with TNF-α. While NO acts as an endogenous neurotransmitter, overproduction of NO can promote neuronal cell death either by direct interaction with cell surface N-methyl-D-aspartate receptors or via interaction with superoxide to produce toxic levels of peroxynitrite. Production of NO by activated microglia and astrocytes is postulated to be the primary source of NO during HIV-associated brain disease. Elevated levels of nitrite in spinal fluid and upregulation of mRNA encoding for iNOS in brain biopsies have been positively correlated with AIDS-related dementia. As a result of these conflicting consequences of NO, it is difficult to conclude whether the suppression of NO mediated by cannabinoids plays a positive, negative or mixed role in the pathogenesis of HIV and AIDS-related diseases.[27]

Coffey and associates[123,124] were the first investigators to demonstrate that THC inhibits production of NO by mouse macrophages both *in vivo* and *in vitro*. Peritoneal macrophages isolated from THC-treated mice produced 50% less NO than cells recovered from control animals when induction *ex vivo* with LPS and IFN-γ. Similarly, NO production was inhibited in mouse macrophages exposed to THC *in vitro* at concentrations ranging from 0.5 to 7 μg/ml. The inhibition of NO was concentration dependent and maximal if THC was added prior to the addition of the inducing agents LPS and IFN-γ. In another report, Jeon and coworkers[125] found that THC inhibited LPS-activation of gene expression for iNOS, as well as NO production, in murine macrophages. Their results suggest that inhibition of cAMP by inhibitory G-protein-coupled cannabinoid receptors attenuates the activation of NF-κB binding protein, which is necessary for the activation of the iNOS gene. Two additional studies using murine cells corroborate these earlier findings.[103,126] Chang and colleagues[103] investigated the pharmacological actions of cannabinoids in the production of NO, IL-6, and PGE$_2$ in a mouse macrophage cell line. They found that both THC and anandamide suppressed LPS-induced production of NO and IL-6 in a concentration-dependent manner. Finally, in a report suggesting that the ability of cannabinoids to affect NO synthesis may lead to biologic effects apart from modulation of macrophage function, Molina-Holgado and coworkers[126] reported that cannabinoids inhibited LPS-induced NO release in primary mouse astrocyte cultures. Specifically, LPS-mediated activation of primary mouse astrocyte cultures resulted in a marked increase in NO release and this effect was abrogated by co-incubation with cannabinoid agonists anandamide and CP-55940.

As just described, much of the available data concerning the impact of cannabinoids on the production of NO and its role as an effecter molecule have been derived from rodent models. The capacity for human inflammatory cells to produce NO and the role of NO as an antimicrobial defense mechanism in humans is more controversial.[127,128] The conditions that usually induce NO in rodent cells often fail to stimulate NO production from human mononuclear cells.[128,129] However, human mononuclear phagocytes have been shown to express iNOS and/or use NO as an antimicrobial effector molecule in some circumstances.[130–132] Recently, in a study utilizing alveolar macrophages (AM) recovered from the lungs of otherwise healthy nonsmokers (NS), smokers of tobacco (TS) and smokers of marijuana (MS), Shay and associates[133] evaluated the capacity for human macrophages to produce NO, the role of this molecule in mediating antimicrobial activity, and the potential for *in vivo* exposure to THC to impair NO production and antimicrobial killing. This work confirmed a significant role for NO as an antibacterial effector used by AM and the presence of impaired host defense in marijuana smokers that was directly related to an inability of their macrophages to produce NO.[133]

In human AM collected from NS and TS, killing of *Staphylococcus aureus* (*S. aureus*) was highly associated with the production of nitrite and induction of iNOS mRNA.[133] Inhibition of iNOS with N^G-monomethyl-L-arginine monoacetate (NGMMA) abrogated the majority of bacterial killing. In contrast, AM from heavy MS failed to express iNOS mRNA or produce NO when co-cultured with *S. aureus*. The functional outcome of these deficits was a significant impairment in antibacterial killing. As already commented on, THC impairs the capacity for T cells and macrophages to produce proinflammatory cytokines that are centrally involved in inducing expression of the iNOS gene, including IFN-γ and TNF-α.[134] The role of cytokine priming was therefore evaluated by adding exogenous IFN-γ or GM-CSF to the AM collected from the lungs of MS. Treatment with either of these inflammatory cytokines restored both NO production and antibacterial activity. Further, this enhancement in antibacterial response was inhibited by NGMMA. This work strongly suggests that marijuana smoking and chronic THC exposure can significantly impair NO production and antibacterial defenses in human AM and that these effects may involve THC-associated impairments in the production of inflammatory cytokines.

Despite evidence that cannabinoids suppress NO production in macrophages, and that this deficit is associated with reduced antimicrobial defenses, there is still the possibility that a net reduction in NO might be beneficial in certain phases of HIV infection. In a test of this hypothesis, Esposito and colleagues[25] studied the effect of cannabinoid CB1 and CB2 receptor agonists on the release of NO and cell toxicity induced by HIV-1 Tat protein in rat glioma cells. They found that the endocannabinoid system protected target glioma cells from Tat-induced overproduction of NO and commensurate cell damage. The clinical relevance of these *in vitro* findings is strengthened by the observations of Boven and coworkers[135] who showed that in AIDS–dementia complex, neuronal damage may be the result of interactions between immune activated glial cells and the consequent and simultaneous production of NO and superoxide anion. Although macrophages play a pivotal role in immune responses and can destroy virally infected cells, they can also be infected by HIV and provide a mechanism for the persistence and tissue dissemination of this virus. Apart from any potential beneficial effect that a cannabinoid-mediated reduction in NO could have on virus replication, suppression of macrophage function by cannabinoids might contribute to complications with broader immunologic implications. Quiescent, functionally inactive macrophages that have been infected by HIV can act as viral reservoirs, posing a long-term threat for continued dissemination of the virus. Additionally, we have shown that diminished NO production in human alveolar macrophages correlates with impaired antibacterial function in marijuana smokers,[133,134] suggesting that an inadequate NO response may contribute to a higher incidence of opportunistic infections in HIV-infected individuals. Finally, there is the potential for cannabinoids to affect the antiviral properties of NO. Although the issue of whether NO acts as a inhibitor of viral replication in HIV disease remains controversial, NO clearly has antiviral properties as manifested in its ability to inactivate enzymes necessary for viral replication.[115,116] Thus, it remains to be seen whether cannabinoid-mediated immune suppression can selectively prevent the harmful effects associated with the overproduction of NO without impairing host defenses and potentiating the harmful consequences of HIV or opportunistic infections. Additional research addressing this important issue is needed.

5. USING THE huPBL-SCID MODEL TO EVALUATE THE *IN VIVO* CONSEQUENCES OF THC ON HIV INFECTION

Epidemiological studies provide suggestive, but indirect, evidence that marijuana potentiates HIV replication and opportunistic infections *in vivo*.[9,11,24] These conclusions are supported by *in vitro* studies demonstrating that cannabinoids directly impair immune function as described above. However, the study of isolated cells in culture cannot address the complex interaction that likely occurs between cannabinoids, HIV, and host responses *in vivo*. In addition, there has been no mechanism for determining whether the potentially beneficial effects of THC on inflammatory cytokines, NO production, and HIV-associated wasting might counterbalance the negative effects of cannabinoids on immune function. Conclusions regarding the impact of cannabinoids on HIV are further complicated by the recent use of highly active antiretroviral therapy (HAART) as a standard of care for HIV patients.[136] It is possible that concurrent administration of HAART and marijuana might effectively prevent the negative effects of THC on viral replication while allowing it to protect against cell injury and wasting. In order to address some of these important issues, we recently established a murine model in which drugs of abuse can be tested for their impact on the infectivity and replication of HIV in human cells *in vivo*.[19] This model employs the human lymphocyte/SCID (huPBL/SCID) mouse developed by Mosier and colleagues.[137] The huPBL/SCID mouse is constructed by transplanting human peripheral blood leukocytes (PBLs) or cord blood cells, both targets for HIV infection, into the peritoneal cavity of SCID mice. Since these mice lack mature B- and T cells, the transplanted human cells are not rejected and can engraft. Various strains of HIV can then be introduced, resulting in the infection of human cells *in vivo* and depletion of the CD4+ population within 2 weeks of exposure to virus. This system has proved useful for assessing viral pathogenesis and we are currently using the huPBL/SCID mouse to analyze the *in vivo* effects of cannabinoids on HIV infectivity, replication, and pathobiology.

To assess the impact of THC on HIV replication *in vivo*, SCID mice were implanted with human PBL and infected 12–13 days later with a functional HIV-reporter construct, NL-r-HSAs.[19,138] This construct, which expresses mouse heat stable antigen (murine CD24) on the surface of HIV-infected cells, allows for rapid and reproducible detection of infected human cells by flow cytometry. Hybrid huPBL/SCID mice that have been infected with HIV can then be treated with other agents, such as THC or cocaine, and evaluated for the impact of these agents on HIV infection (as measured by expression of murine CD24, CD4 counts, CD4/CD8 ratio). We found that THC, even in the absence of HIV, significantly decreased the number of CD4 positive T cells recovered at 1–2 weeks after huPBL engraftment (Fig. 1A). This effect might be similar to the thymic atrophy and decrease in splenic cellularity observed when control mice are treated with THC and suggests that THC promotes both a reduction in proliferation and apoptosis *in vivo*.[82] Despite this diminution in HIV target cells, daily exposure to THC in the huPBL/SCID model resulted in a 2.2-fold increase in HIV-infected cells harvested at 7 days post-infection (Fig. 1B). These preliminary studies utilizing the huPBL/SCID model indicate that THC enhances HIV replication and the subsequent destruction of human immune cells *in vivo*. This model provides an excellent opportunity to evaluate the role of THC-induced cytokine changes, NO production, HAART, and other potential mechanisms/modulators on the complex interaction between THC, host immunity, and viral pathogenesis.

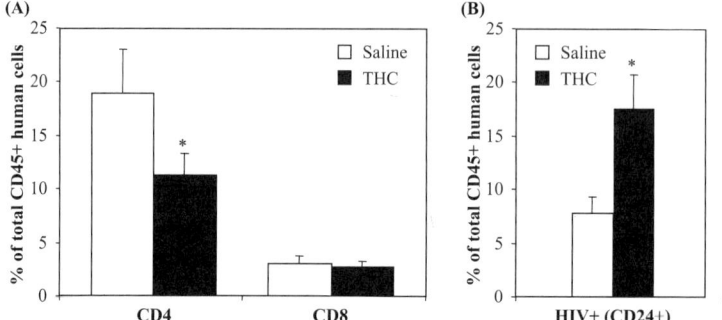

FIGURE 1. (A) THC, in the absence of infection by HIV, significantly decreased the number of CD4+ T cells and the CD4 : CD8 ratio in the huPBL/SCID model. SCID mice were engrafted with 1-2 × 10⁷ human PBL and after 12 days treated for 10 days with administration of either THC (10 mg/kg) or saline by intraperitoneal injection for 4 consecutive days each week. Peritoneal lavage was performed on day 21 to recover implanted cells and the percentage of CD4+ and CD8+ cells determined by FACS analysis after gating for cells expressing the human CD45 antigen. Values represent means ±SD, $n = 15$ per group. *$p < 0.01$. (B) Infection with HIV is augmented by systemic administration of THC. SCID mice were implanted with human PBL, infected with the HIV reporter construct NL-r-HSAs that expresses murine CD24 antigen on infected cells, and treated daily with THC (5 mg/kg) or saline for 7 days. Peritoneal lavage was then performed and recovered cells evaluated by FACS analysis for the presence of HIV-infected human PBL (% of CD45+ cells expressing CD24+). *In vivo* exposure to THC resulted in a 2.2-fold increase in HIV-positive cells. Values represent means ±SD, $n = 15$ per group. *$p < 0.01$.

6. SUMMARY

The pathogenesis of AIDS is a complex and prolonged process that can be altered by a variety of cofactors, including the abuse of illicit drugs. The exact mechanisms by which THC facilitates this disease are yet to be proven, but likely include a combination of increased risk due to drug-related social behaviors, a wide-ranging capacity for THC to suppress host immunity, and effects on the infectivity and replication of HIV. The huPBL/SCID model provides a system in which the role of different cannabinoid receptors, cytokines, and other mechanisms can be investigated for their impact on HIV infectivity, replication, and toxicity *in vivo*. Further studies are also needed to clarify the effects of THC on human cells *in vitro*, to examine the impact of real-life exposure to marijuana on the immune system of chronic abusers, and to investigate the mechanisms by which these effects interact with the pathogenesis of HIV and AIDS in patients. At the same time, clinical trials are evaluating whether the combination of HAART and THC will allow potentially beneficial effects of cannabinoids to be realized while limiting the impact of these drugs on the progression of HIV and AIDS.[23] Considering the frequent use of marijuana and THC by patients with HIV, further research into the biological links between cannabinoids with the pathogenesis of HIV should be a high priority.

ACKNOWLEDGMENTS. Financial support was provided by the National Institutes of Health/National Institute on Drug Abuse (grants RO1 DA08254 and R37 DA03018).

REFERENCES

1. Feng, Y., Broder, C. C., Kennedy, P. E., and Berger, E. A. (1996). HIV-1 entry cofactor: Functional cDNA cloning of a seven-transmembrane, G protein-coupled receptor. *Science* **272:**872–877.
2. Peterson, P. K., Gekker, G., Chao, C. C., Hu, S., Edelman, C., Balfour, H. H. Jr. *et al.* (1992). Human cytomegalovirus-stimulated peripheral blood mononuclear cells induce HIV-1 replication via a tumor necrosis factor-a-mediated mechanism. *J. Clin. Invest.* **89:**574–580.
3. Strathdee, S. A., Hoggs, R. S., O'Shaughnessy, M. V., Montaner, J. S., and Schechter, M. T. (1996). A decade of research on the natural history of HIV infection: Part 2. Cofactors. *Clin. Invest. Med.* **19:**121–130.
4. Timbo, B. B. and Tollefson, L. (1994). Nutrition: A cofactor in HIV disease. *J. Am. Diabet. Assoc.* **94:**1018–1022.
5. Whalen, C., Horsburgh, C. R., Hom, D., Lahart, C., Simberkoff, M., and Ellner, J. (1995). Accelerated course of human immunodeficiency virus infection after tuberculosis. *Am. J. Respir. Crit. Care Med.* **151:**129–135.
6. Baldwin, G. C., Roth, M. D., and Tashkin, D. P. (1998). Acute and chronic effects of cocaine on the immune system and the possible link to AIDS. *J. Neuroimmunol.* **83:**133–138.
7. Dingle, G. A. and Oei, T. P. (1997). Is alcohol a cofactor of HIV and AIDS? Evidence from immunological and behavioral studies. *Psychol. Bull.* **122:**56–71.
8. Donahoe, R. M. and Vlahov, D. (1998). Opiates as potential cofactors in progression of HIV-1 infections to AIDS. *J. Neuroimmunol.* **8:**77–87.

9. Newell, G. R., Mansell, P. W., Wilson, M. B., Lynch, H. K., Spitz, M. R., and Hersh, E. M. (1985). Risk factor analysis among men referred for possible acquired immune deficiency syndrome. *Prev. Med.* **14**:81–91.

10. Soderberg, L. S. (1998). Immunomodulation by nitrite inhalants may predispose abusers to AIDS and Kaposi's Sarcoma. *J. Neuroimmunol.* **83**:157–161.

11. Tindall, B., Philpot, C. R., Cooper, D. A., Gold, J., Donovan, B., Penny, R. *et al.* (1988). The Sydney AIDS project: Development of acquired immunodeficiency syndrome in a group of HIV seropositive homosexual men. *Aust. NZ J. Med.* **18**:8–15.

12. Wang, Y. and Watson, R. R. (1995). Is alcohol consumption a cofactor in the development of acquired immunodeficiency syndrome? *Alcohol* **12**:105–109.

13. Rasch, R. F., Weisen, C. A., MacDonald, B., Wechsberg, W. M., Perritt, R., and Dennis, M. L. (2000). Patterns of HIV risk and alcohol use among African-American crack abusers. *Drug Alcohol Depend.* **58**:259–266.

14. Wang, M. Q., Collins, C. B., Kohler, C. L., DiClemente, R. J., and Wingood, G. (2000). Drug use and HIV risk-related sex behaviors: A street outreach study of black adults. *South Med. J.* **9**:186–190.

15. Klein, T. W., Newton, C. A., Nakachi, N., and Friedman, H. (2000). Delta 9-tetrahydro-cannabinol treatment suppresses immunity and early IFN-gamma, IL-12, and IL-12 receptor beta2 responses to *Legionella pneumophila* infection. *J. Immunol.* **164**:6461–6466.

16. Klein, T. W., Lane, B., Newton, C. A., and Friedman, H. (2000). The cannabinoid system and cytokine network. *Proc. Soc. Exp. Med. Biol.* **225**:1–8.

17. Pillai, R., Nair, B. S., and Watson, R. R. (1991). AIDS, drugs of abuse and the immune system: A complex immunotoxicological network. *Arch. Toxicol.* **65**:609–617.

18. Peterson, P. K., Gekker, G., Chao, C. C., Schut, R., Molitor, T. W., and Balfour, H. H., Jr. (1991). Cocaine potentiates HIV-1 replication in human peripheral blood mononuclear cell cocultures. Involvement of transforming growth factor beta. *J. Immunol.* **146**:81–84.

19. Roth, M. D., Tashkin, D. P., Choi, R., Jamieson, D. D., Zack, J. A., and Baldwin, G. C. (2002). Cocaine enhances HIV replication in a HuPBL-SCID mouse model. *J. Infect. Dis.* **185**:701–705.

20. Roth, M. D., Baldwin, G. C., and Tashkin, D. P. (2002). Effects of delta-9-tetrahydro-cannabinol on human immune function and host defense. *Chem. Phys. Lipids* **121**:229–239.

21. Weber, A. E., Chan, K., George, C., Hogg, R. S., Remis, R. S., Martindale, S. *et al.* (2001). Risk factors associated with HIV infection among young gay and bisexual men in Canada. *J. Acquir. Immune Defic. Syndr.* **28**:81–88.

22. Beal, J. E., Olson, R., Lefkowitz, L., Laubenstein, L., Bellman, P., Yangco, B. *et al.* (1997). Long-term efficacy and safety of dronabinol for acquired immunodeficiency syndrome-associated anorexia. *J. Pain. Symptom Manage.* **14**:7–14.

23. Bredt, B. M., Higuera-Alhino, D., Shade, S. B., Hebert, S. J., McCune, J. M., and Abrams, D. I. (2002). Short-term effects of cannabinoids on immune phenotype and function in HIV-1-infected patients. *J. Clin. Pharmacol.* **42**(Suppl):82S–89S.

24. Caiaffa, W. T., Vlahov, D., Graham, N. M. H., Astemborski, J., Solomon, L., Nelson, K. E. *et al.* (1994). Drug smoking, *Pneumocystis carinii* pneumonia, and immunosuppression increase risk of bacterial pneumonia in human immunodeficiency virus-seropositive injection drug users. *Am. J. Respir. Crit. Care Med.* **150**:1493–1498.

25. Esposito, G., Ligresti, A., Izzo, A. A., Bisogno, T., Ruvo, M., Di Rosa, M., *et al.* (2002). The endocannabinoid system protects rat glioma cells against HIV-1 Tat protein-induced cytotoxicity. Mechanism and regulation. *J. Biol. Chem.* **277**:50348–50354.

26. Klegeris, A., Bissonnette, C. J., and McGeer, P. L. (2003). Reduction of human monocyte cell neurotoxicity and cytokine secretion by ligands of the cannabinoid-type CB2 receptor. *Br. J. Pharmacol.* **139**:775–786.

27. Torre, D., Pugliese, A., and Speranza, F. (2002). Role of nitric oxide in HIV-1 infection: Friend or foe? *Lancet Infect. Dis.* **2**:273–280.

28. Harvey, D. J. (1984). Chemistry, metabolism, and pharmacokinetics of the cannabinoids. In G. G. Nahas (ed.), *Marijuana in Science and Medicine*, Raven Press, New York, pp. 37–107.

29. Novotny, M., Merli, F., Weisler, D., Fencl, M., and Saeed, T. (1982). Fractionation and capillary gas chromatographic–mass spectrometric characterization of the neutral components in marijuana and tobacco smoke condensates. *J. Chomatogr.* **238:**141–150.

30. Roth, M. D., Marquez-Magallanes, J. A., Yuan, M., Sun, W., Tashkin, D. P., and Hankinson, O. (2001). Induction and regulation of the carcinogen-metabolizing enzyme, CYP1A1, by marijuana smoke and Δ^9-Tetrahydrocannabinol. *Am. J. Respir. Cell Mol. Biol.* **24:**1–6.

31. Gaoni, Y. and Mechoulam, R. (1964). Hashish III: Isolation, structure and partial synthesis of an active constituent of hashish. *J. Am. Chem. Soc.* **86:**1646–1647.

32. Devane, W. A., Hanus, L., Breuer, A., Pertwee, R. G., Stevenson, L. A., Griffin, G. *et al.* (1992). Isolation and structure of a brain constituent that binds to the cannabinoid receptor. *Science* **258:**1946–1949.

33. Di Marzo, V., Bisogno, T., De Petrocellis, L., Melck, D., and Martin, B. R. (1999). Cannabimimetic fatty acid derivatives: The anandamide family and other endocannabinoids. *Curr. Med. Chem.* **6:**721–744.

34. Hanus, L., Abu-Lafi, S., Fride, E., Breuer, A., Vogel, Z., Shalev, D. E. *et al.* (2001). 2-Arachidonyl glyceryl ether, an endogenous agonist of the cannabinoid CB1 receptor. *Proc. Natl. Acad. Sci. USA* **98:**3662–3665.

35. Lee, M., Yang, K. H., and Kaminski, N. E. (1995). Effects of putative cannabinoid receptor ligands, anandamide and 2-arachidonyl-glycerol, on immune function in B6C3F1 mouse splenocytes. *J. Pharmacol. Exp. Ther.* **275:**529–536.

36. Gerard, C. M., Mollereau, C., Vassart, G., and Parmentier, M. (1991). Molecular cloning of a human cannabinoid receptor which is also expressed in testis. *Biochem. J.* **279:**129–134.

37. Matsuda, L. A., Lolait, S. J., Brownstein, M. J., Young, A. C., and Bonner, T. I. (1990). Structure of a cannabinoid receptor and functional expression of the cloned cDNA. *Nature* **346:**561–564.

38. Bouaboula, M., Rinaldi, M., Carayon, P., Carillon, C., Delpech, B., Shire, D. *et al.* (1993). Cannabinoid-receptor expression in human leukocytes. *Eur. J. Biochem.* **214:**173–180.

39. Galiegue, S., Mary, S., Marchand, J., Dussossoy, D., Carriere, D., Carayon, P. *et al.* (1995). Expression of central and peripheral cannabinoid receptors in human immune tissues and leukocyte subpopulations. *Eur. J. Biochem.* **232:**54–61.

40. Ledent, C., Valverde, O., Cossu, G., Petitet, F., Aubert, J. F., Beslot, F. *et al.* (1999). Unresponsiveness to cannabinoids and reduced addictive effects of opiates in CB1 receptor knockout mice. *Science* **283:**401–404.

41. Huestis, M. A., Gorelick, D. A., Heishman, S. J., Preston, K. L., Nelson, R. A., Moolchan, E. T. *et al.* (2001). Blockade of effects of smoked marijuana by the CB1-selective cannabinoid receptor antagonist SR141716. *Arch. Gen. Psychiatry* **58:**322–328.

42. Munro, S., Thomas, K. L., and Abu-Shaar, M. (1993). Molecular characterization of a peripheral receptor for cannabinoids. *Nature* **365:**61–65.

43. Nong, L., Newton, C., Cheng, Q., Friedman, H., Roth, M. D., and Klein, T. W. (2002). Altered cannabinoid receptor mRNA expression in peripheral blood mononuclear cells from marijuana smokers. *J. Neuroimmunol.* **127:**169–176.

44. Marchand, J., Bord, A., Penarier, G., Laure, F., Carayon, P., and Casellas, P. (1999). Quantitative method to determine mRNA levels by reverse transcriptase-polymerase chain reaction from leukocyte subsets purified by fluorescence-activated cell sorting: Application to peripheral cannabinoid receptors. *Cytometry* **35:**227–234.

45. Howlett, A. C. (1995). Pharmacology of cannabinoid receptors. *Ann. Rev. Pharmacol. Toxicol.* **35:**607–634.

46. Berdyshev, E. V. (2000). Cannabinoid receptors and the regulation of immune response. *Chem. Phys. Lipids* **108:**169–190.

47. Kaminski, N. E. (1998). Regulation of the cAMP cascade, gene expression and immune function by cannabinoid receptors. *J. Neuroimmunol.* **83:**124–132.

48. Fride, E. (2002). Endocannabinoids in the central nervous system—an overview. *Prostaglandins Leukot. Essent. Fatty Acids* **66:**221–233.

49.   Howlett, A. C., Barth, F., Bonner, T. I., Cabral, G., Casellas, P., Devane, W. A. *et al.* (2002). International Union of Pharmacology. XXVII. Classification of cannabinoid receptors. *Pharmacol. Rev.* **54**:161–202.

50.   Klein, T. W., Newton, C., and Friedman, H. (1998). Cannabinoid receptors and immunity. *Immunol. Today* **19**:373–381.

51.   Palmer S. L., Thakur, G. A., and Makriyannis, A. (2002). Cannabinergic ligands. *Chem. Phys. Lipids* **121**:3–19.

52.   Breen, E. C., Rezai, A. R., Nakajima, K., Hirano, T., Beall, G. N., Mitsuyasu, R. T. *et al.* (1990). Elevated levels of interleukin 6 (IL-6) are associated with human immunodeficiency virus (HIV) infection. *J. Immunol.* **144**:480–484.

53.   Odeh, M. (1990). The role of tumour necrosis factor-alpha in acquired immunodeficiency syndrome. *J. Intern. Med.* **228**:549–556.

54.   Breen, E. C. (2002). Pro- and anti-inflammatory cytokines in human immunodeficiency virus infection and acquired immunodeficiency syndrome. *Pharmacol. Ther.* **95**:295–304.

55.   Clerici, M., Wynn, T. A., Berzofsky, J. A., Blatt, S. P., Hendrix, C. W., Sher, A. *et al.* (1994). Role of interleukin-10 in T helper cell dysfunction in asymptomatic individuals infected with the human immunodeficiency virus. *J. Clin. Invest.* **93**:768–775.

56.   Kekow, J., Wachsman, W., McCutchan, J. A., Cronin, M., Carson, D. A., and Lotz, M. (1990). Transforming growth factor beta and noncytopathic mechanisms of immunodeficiency in human immunodeficiency virus infection. *Proc. Natl. Acad. Sci. USA* **87**:8321–8325.

57.   Lane, H. C., Masur, H., Edgar, L. C., Whalen, G., Rook, A. H., and Fauci, A. S. (1983). Abnormalities of B-cell activation and immunoregulation in patients with the acquired immunodeficiency syndrome. *N. Engl. J. Med.* **309**:453–458.

58.   Chehimi, J., Starr, S. E., Frank, I., D'Andrea, A., Ma, X., MacGregor, R. R. *et al.* (1994). Impaired interleukin 12 production in human immunodeficiency virus-infected patients. *J. Exp. Med.* **179**:1361–1366.

59.   Ma, X. and Montaner, L. J. (2000). Proinflammatory response and IL-12 expression in HIV-1 infection. *J. Leukoc. Biol.* **68**:383–390.

60.   Broder, C. C. and Collman, R. G. (1997). Chemokine receptors and HIV. *J. Leukoc. Biol.* **62**:20–29.

61.   Kitchen, S., Korin, Y., Roth, M. D., Landay, A., and Zack, J. A. (1998). Costimulation of naive CD8+ lymphocytes induces CD4 expression and allows HIV-1 infection. *J. Virol.* **72**:9054–9060.

62.   Steinman, R. M., Granelli-Piperno, A., Pope, M., Trumpfheller, C., Ignatius, R., Arrode, G. *et al.* (2003). The interaction of immunodeficiency viruses with dendritic cells. *Curr. Top. Microbiol. Immunol.* **276**:1–30.

63.   Weissman, D. and Fauci, A. S. (1997). Role of dendritic cells in immunopathogenesis of human immunodeficiency virus infection. *Clin. Microbiol. Rev.* **10**:358–367.

64.   Ruegg, C. L. and Engleman, E. G. (1990). Impaired immunity in AIDS. The mechanisms responsible and the potential reversal by antiviral therapy. *Ann NY Acad. Sci.* **616**:307–317.

65.   Cabral, G. A. and Dove Pettit, D. A. (1998). Drugs and immunity: Cannabinoids and their role in decreased resistance to infectious disease. *J. Neuroimmunol.* **83**:116–123.

66.   Klein, T. W., Newton, C. A., Larson, K., Lu, L., Perkins, I., Nong, L. *et al.* (2003). The cannabinoid system and immune modulation. *J. Leukoc. Biol.* **74**:486–496.

67.   Salzet, M., Breton, C., Bisogno, T., and Di Marzo, V. (2000). Comparative biology of the endocannabinoid system possible role in the immune response. *Eur. J. Biochem.* **267**:4917–4927.

68.   Nahas, G. G., Zagury, D., Schwartz, I. W., and Nagel, M. D. (1973). Evidence for the possible immunogenicity of delta 9-tetrahydrocannabinol (THC) in rodents. *Nature* **243**:407–408.

69.   Nahas, G. G., Suciu-Foca, N., Armand, J.-P., and Morishima, A. (1974). Inhibition of cellular mediated immunity in marijuana smokers. *Science* **183**:419–420.

70.   Newton, C. A., Klein, T. W., and Friedman, H. (1994). Secondary immunity to *Legionella pneumophila* and Th1 activity are suppressed by delta-9-tetrahydrocannabinol injection. *Infect. Immun.* **62**:4015–4020.

71. Zhu, L. X., Sharma, S., Stolina, M., Gardner, B., Roth, M. D., Tashkin, D. P. *et al.* (2000). Delta-9-tetrahydrocannabinol inhibits antitumor immunity by a CB2 receptor-mediated, cytokine-dependent pathway. *J. Immunol.* **165**:373–380.

72. Mosmann, T. R. and Moore, K. W. (1991). The role of IL-10 in cross-regulation of T_H1 and T_H2 responses. *Immunol. Today* **12**:A49–A53.

73. Huang, M., Wang, J., Lee, P., Sharma, S., Mao, J. T., Meissner, H. *et al.* (1995). Human non-small cell lung cancer cells express a type 2 cytokine pattern. *Cancer Res.* **55**:3847–3853.

74. Sogn, J. A. (1998). Tumor immunology: The glass is half full. *Immunity* **9**:757–763.

75. Gardner, B., Zhu, L. X., Sharma, S., Liu, Q., Makriyannis, A., Tashkin, D. P. *et al.* (2002). Autocrine and paracrine regulation of lymphocyte CB2 receptor expression by TGF-β. *Biochem. Biophys. Res. Commun.* **290**:91–96.

76. Yuan, M., Kiertscher, S. M., Cheng, Q., Zoumalan, R., Tashkin, D. P., and Roth, M. D. (2002). Δ^9-Tetrahydrocannabinol regulates Th1/Th2 cytokine balance in activated human T-cells. *J. Neuroimmunol.* **133**:124–131.

77. Fischer-Stenger, K., Dove Pettit, D. A., and Cabral, G. A. (1993). Δ^9-Tetrahydrocannabinol inhibition of tumor necrosis factor-α suppression of post-translational events. *J. Pharmacol. Exp. Ther.* **267**:1558–1565.

78. Klein, T. W., Newton, C., Widen, R., and Friedman, H. (1993). Δ^9-Tetrahydrocannabinol injection induces cytokine-mediated mortality of mice infected with *Legionella pneumophila. J. Pharmacol. Exp. Ther.* **267**:635–640.

79. Zhu, W., Newton, C., Daaka, Y., Friedman, H., and Klein, T. W. (1994). Δ^9-Tetrahydrocannabinol enhances the secretion of interleukin-1 from endotoxin-stimulated macrophage. *J. Pharmacol. Exp. Ther.* **270**:1334–1339.

80. Carayon, P., Marchand, J., Dussossoy, D., Derocq, J. M., Jbilo, O., Bord, A. *et al.* (1998). Modulation and functional involvement of CB2 peripheral cannabinoid receptors during B-cell differentiation. *Blood* **92**:3605–3615.

81. Schatz, A. R., Koh, W. S., and Kaminski, N. E. (1993). Delta 9-tetrahydrocannabinol selectively inhibits T-cell dependent human immune responses through direct inhibition of accessory cell function. *Immunopharmacology* **26**:129–137.

82. McKallip, R. J., Lombard, C., Fisher, M., Martin, B. R., Ryu, S., Grant, S. *et al.* (2002). Targeting CB2 cannabinoid receptors as a novel therapy to treat malignant lymphoblastic disease. *Blood* **100**:627–634.

83. Schwarz, H., Blanco, F. J., and Lotz, M. (1994). Anandamide, an endogenous cannabinoid receptor agonist inhibits lymphocyte proliferation and induces apoptosis. *J. Neuroimmunol.* **55**:107–115.

84. Zhu, W., Friedman, H., and Klein, T. W. (1998). Delta 9-tetrahydrocannabinol induces apoptosis in macrophages and lymphocytes: Involvement of Bcl-2 and caspase-1. *J. Pharmacol. Exp. Ther.* **286**:1103–1109.

85. Bennasser, Y. and Bahraoui, E. (2002). HIV-1 Tat protein induces interleukin-10 in human peripheral blood monocytes: Involvement of protein kinase C-betaII and -delta. *FASEB J.* **16**:546–554.

86. Rasty, S., Thatikunta, P., Gordon, J., Khalili, K., Amini, S., and Glorioso, J. C. (1996). Human immunodeficiency virus tat gene transfer to the murine central nervous system using a replication-defective herpes simplex virus vector stimulates transforming growth factor beta 1 gene expression. *Proc. Natl. Acad. Sci. USA* **93**:6073–6078.

87. Taoufik, Y., Lantz, O., Wallon, C., Charles, A., Dussaix, E., and Delfraissy, J. F. (1997). Human immunodeficiency virus gp120 inhibits interleukin-12 secretion by human monocytes: An indirect interleukin-10-mediated effect. *Blood* **89**:2842–2848.

88. Wang, J., Guan, E., Roderiquez, G., and Norcross, M. A. (2001). Synergistic induction of apoptosis in primary CD4(+) T cells by macrophage-tropic HIV-1 and TGF-beta1. *J. Immunol.* **167**:3360–3366.

89. Bodi, I., Kimmel, P. L., Abraham, A. A., Svetkey, L. P., Klotman, P. E., and Kopp, J. B. (1997). Renal TGF-beta in HIV-associated kidney diseases. *Kidney Int.* **51**:1568–1577.

90. Lima, R. G., Van Weyenbergh, J., Saraiva, E. M., Barral-Netto, M., Galvao-Castro, B., and Bou-Habib, D. C. (2002). The replication of human immunodeficiency virus type 1 in macrophages is enhanced after phagocytosis of apoptotic cells. *J. Infect. Dis.* **185:**1561–1566.

91. Lotz, M. and Seth, P. (1993). TGF beta and HIV infection. *Ann. NY Acad Sci.* **685:**501–511.

92. McKallip, R. J., Lombard, C., Martin, B. R., Nagarkatti, M., and Nagarkatti, P. S. (2002). Delta(9)-tetrahydrocannabinol-induced apoptosis in the thymus and spleen as a mechanism of immunosuppression *in vitro* and *in vivo*. *Blood* **100:**627–634.

93. Ancuta, P., Bakri, Y., Chomont, N., Hocini, H., Gabuzda, D., and Haeffner-Cavaillon, N. (2001). Opposite effects of IL-10 on the ability of dendritic cells and macrophages to replicate primary CXCR4-dependent HIV-1 strains. *J. Immunol.* **166:**4244–4253.

94. Houle, M., Thivierge, M., Le Gouill, C., Stankova, J., and Rola-Pleszczynski, M. (1999). IL-10 up-regulates CCR5 gene expression in human monocytes. *Inflammation* **23:**241–251.

95. Weissman, D., Poli, G., and Fauci, A. S. (1995). IL-10 synergizes with multiple cytokines in enhancing HIV production in cells of monocytic lineage. *J. Acquir. Immune Defic. Syndr. Hum. Retrovirol.* **9:**442–449.

96. Baronzio, G., Zambelli, A., Comi, D., Barlocco, A., Baronzio, A., Marchesi, P. *et al.* (1999). Proinflammatory and regulatory cytokine levels in AIDS cachexia. *In Vivo,* **13:**499–502.

97. Jimenez, J. L., Gonzalez-Nicolas, J., Alvarez, S., Fresno, M., and Munoz-Fernandez, M. A. (2001). Regulation of human immunodeficiency virus type 1 replication in human T lymphocytes by nitric oxide. *J. Virol.* **75:**4655–4663.

98. Brendt, D. S. and Snyder, S. H. (1994). Nitric oxide: A physiologic messenger molecule. *Ann. Rev. Biochem.* **63:**175–195.

99. Mayer, B. (2000). Nitric oxide. In B. Mayer, P. Cuatrecasas, K. Starke, G. V. Born, and P. Taylor (eds), *Handbook of Experimental Pharmacology, Volume: Nitric Oxide,* Springer-Verlag, New York, pp. 1–667.

100. Moncada, S., Palmer, R. M., and Higgs, E. A. (1991). Nitric oxide physiology, pathophysiology, and pharmacology. *Pharmacol. Rev.* **43:**109–142.

101. Bogdan, C. (2000). The function of nitric oxide in the immune system. In B. Mayer, P. Cuatrecasas, K. Starke, G. V. Born, and P. Taylor (eds), *Handbook of Experimental Pharmacology, Volume: Nitric Oxide,* Springer-Verlag, New York, pp. 443–492.

102. Bogdan, C. (2001). Nitric oxide and the immune response. *Nat. Immunol.* **2:**907–916.

103. Chang, Y.-H., Lee, S. T., and Lin, W.-W. (2001). Effects of cannabinoids on LPS-stimulated inflammatory mediator release from macrophages: Involvement of eicosanoids. *J. Cell. Biochem.* **81:**715–723.

104. Stuehr, D. J. and Marletta, M. A. (1987). Induction of nitrite/nitrate synthesis in murine macrophages by BCG infection, lymphokines or interferon-γ. *J. Immunol.* **139:**518–525.

105. Akarid, K., Sinet, M., Desforges, B., and Gougerot-Pocidalo, M. A. (1995). Inhibitory effect of nitric oxide on the replication of murine retrovirus *in vitro* and *in vivo. J. Virol.* **69:**7001–7005.

106. Croen, K. D. (1993). Evidence for antiviral effect of nitric oxide. Inhibition of herpes simplex virus type 1 replication. *J. Clin. Invest.* **91:**2446–2452.

107. Karupiah, G., Xie, Q., Buller, R. M. L., Nathan, C., Duarte, C., and MacMicking, J. D. (1993). Inhibition of viral replication by interferon-gamma induced nitric oxide synthase. *Science* **61:**1445–1448.

108. Mannick, J. B. (1995). The antiviral role of nitric oxide. *Res. Immunol.* **146:**693–697.

109. Blond, D., Cheret, A., Raoul, H., LeGrand, R., Caufour, P., Theodoro, F. *et al.* (1998). Nitric oxide synthesis during acute SIV mac251 infection of macaques. *Res. Virol.* **149:**75–86.

110. Groeneveld, P. H., Kroon, F. P., Nibbering, P. H., Bruisten, S. M., van Swieten, P., and van Furth, R. (1996). Increased production of nitric oxide correlates with viral load and activation of mononuclear phagocytes in HIV-infected patients. *Scand. J. Infect. Dis.* **28:**341–345.

111. Zangerle, R., Fuchs, D., Reinegger, G., Werner-Felmayer, G., Gallati, H., Wachter, H. *et al.* (1995). Serum nitrite and nitrate in infection with human immunodeficiency virus type-1. *Immunobiology* **193:**59–70.

112. Zhao, M.-L., Kim, M.-O., Morgello, S., and Lee, S. C. (2001). Expression of inducible nitric oxide synthase, interleukin-1 and caspase-1 in HIV-1 encephalitis. *J. Neuroimmun.* **115:**182–191.

113. Chen, F., Lu, Y., Castranova, V., Rojanasakul, Y., Miyahare, K., Shizuta, Y. *et al.* (1999). Nitric oxide inhibits HIV tat-induced NFκB activation. *Am. J. Pathol.* **155:**275–284.

114. Mannick, J. B., Stamler, J. S., Teng, E., Simpson, N., Lawrence, J., Jordan, J. *et al.* (1999). Nitric oxide modulates HIV-1 replication. *J. Acquir. Immune. Defic. Syndr.* **22:**1–9.

115. Persichini, T., Colasanti, M., Lauro, G. M., and Ascenzi, P. (1998). Cysteine nitrosylation inactivates the HIV-1 protease. *Biochem. Biophys. Res. Commun.* **250:**575–576.

116. Persichini, T., Colasanti, M., Fraziano, M., Colizzi, V., Ascenzi, P., and Lauro, G. M. (1999). Nitric oxide inhibits HIV-replication in human astrocytoma cells. *Biochem. Biophys. Res. Commun.* **254:**200–202.

117. Persichini, T., Colasanti, M., Fraziano, M., Colizzi, V., Medana, C., Polticelli, F. *et al.* (1999). Nitric oxide inhibits the HIV-1 reverse transcriptase activity. *Biochem. Biophys. Res. Commun.* **258:**624–627.

118. Sekkai, D., Aillet, F., Israel, N., and Lepoivre, M. (1998). Inhibition of NF-κB and HIV-1 long terminal repeat transcriptional activation by inducible nitric oxide synthase 2 activity. *J. Biol. Chem.* **273:**3895–3900.

119. Fehsel, K., Kroncke, K. D., Meyer, K. L., Huber, H., Wahn, V., and Kolb-Bachofen, V. (1995). Nitric oxide induces apoptosis in mouse thymocytes. *J. Immunol.* **155:** 2858–2865.

120. Mossalayi, M. D., Becherel, P. A., and Debre, P. (1999). Critical role of nitric oxide during apoptosis of peripheral blood leukocytes from patients with AIDS. *Mol. Med.* **5:**812–819.

121. Taylor-Robinson, A. W. (1997). Inhibition of IL-2 production by nitric oxide: A novel self-regulatory mechanism for Th1 proliferation. *Immunol. Cell. Biol.* **75:**167–175.

122. Torre, D. and Ferrario, G. (1996). Immunological aspects of nitric oxide in HIV-1 infection. *Med. Hypotheses* **47:**405–407.

123. Coffey, R. G., Snella, E., Johnson, K., and Pross, S. (1996). Inhibition of macrophage nitric oxide production by tetrahydrocannabinol *in vivo* and *in vitro*. *Int. J. Immunopharmacol.* **18:**749–752.

124. Coffey, R. G., Yamamoto, Y., Snella, E., and Pross, S. (1996). Tetrahydrocannabinol inhibition of macrophage nitric oxide production. *Biochem. Pharm.* **52:**743–751.

125. Jeon, Y., Yang, K. H., Pulaski, J. T., and Kaminski, N. E. (1996). Attenuation of inducible nitric oxide synthase gene expression by delta 9-tetrahydrocannabinol is mediated through the inhibition of nuclear factor-kappa B/Rel activation. *Mol. Pharmacol.* **50:**334–341.

126. Molina-Holgado, F., Molina-Holgado, E., Guaza, C., and Rothwell, N. J. (2002). Role of CB1 and CB2 receptors in the inhibitory effects of cannabinoids on lipopolysaccharide-induced nitric oxide release in astrocyte cultures. *J. Neurosci. Res.* **67:**829–836.

127. Albina, J. E. (1995). On the expression of nitric oxide synthase by human macrophages. Why no NO? *J. Leukoc. Biol.* **58:**643–649.

128. Denis, M. (1994). Human monocytes/macrophages: NO or no NO? *J. Leukoc. Biol.* **55:**682–684.

129. Zhang, X., Laubach, V. E., Alley, E. W., Edwards, K. A., Sherman, P. A., Russell, S. W. *et al.* (1996). Transcriptional basis for hyporesponsiveness of the human inducible nitric oxide synthase gene to lipopolysaccharide/interferon-gamma. *J. Leukoc. Biol.* **59:**575–585.

130. Nicholson, S., Bonecini-Almeida Mda, G., Lapa e Silva, J. R., Nathan, C., Xie, Q. W., Mumford, R. *et al.* (1996). Inducible nitric oxide synthase in pulmonary alveolar macrophages from patients with tuberculosis. *J. Exp. Med.* **183:**2293–2302.

131. Nozaki, Y., Hasegawa, Y., Ichiyama, S., Nakashima, I., and Shimokata, K. (1997). Mechanism of nitric oxide-dependent killing of *Mycobacterium bovis* BCG in human alveolar macrophages. *Infect. Immun.* **65:**3644–3647.

132. Wang, C. H., Liu, C. Y., Lin, H. C., Yu, C. T., Chung, K. F., and Kuo, H. P. (1998). Increased exhaled nitric oxide in active pulmonary tuberculosis due to inducible NO synthase upregulation in alveolar macrophages. *Eur. Respir. J.* **11**:809–815.

133. Shay, A. H., Choi, R., Whittaker, K., Salehi, K., Kitchen, C. M. R., Tashkin, D. P. *et al.* (2003). Impairment of antimicrobial activity and nitric oxide production in alveolar macrophages from smokers of marijuana and cocaine. *J. Infect. Dis.* **187**:700–704

134. Baldwin, G. C., Tashkin, D. P., Buckley, D. M., Park, A. N., Dubinett, S. M., and Roth, M. D. (1997). Marijuana and cocaine impair alveolar macrophage function and cytokine production. *Am. J. Respir. Crit. Care Med.* **156**:1606–1613.

135. Boven, L. A., Gomes, L., Hery, C., Gray, F., Verhoef, J., Portegies, P. *et al.* (1999). Increased peroxynitrite activity in AIDS dementia complex: Implications for the neuropathogenesis of HIV-1 infection. *J. Immunol.* **162**:4319–4327.

136. Richman, D. D. (2001). HIV chemotherapy. *Nature* **410**:995–1001.

137. Mosier, D. E., Gulizia, R. J., Baird, S. M., and Wilson, D. B. (1988). Transfer of a functional human immune system to mice with severe combined immunodeficiency. *Nature* **335**:256–259.

138. Jamieson, B. D. and Zack, J. A. (1998). *In vivo* pathogenesis of a human immunodeficiency virus type 1 reporter virus. *J. Virol.* **72**:6520–6526.

3

Cannabinoids and Herpesviruses

JERRY L. BULEN and PETER G. MEDVECZKY

1. INTRODUCTION

The effect of cannabinoids on the immune system has been studied in great detail over the last three decades primarily by *in vitro* systems and in experimental animal models. The overwhelming majority of these studies suggest that the main psychoactive component of marijuana, Δ^9-tetrahydracannabinol (THC), has negative effects on immunity.[1,2] This review will summarize the current state of knowledge on herpesvirus infection and cannabinoids. This topic was selected because after primary infection, herpesviruses cause lifelong latent infection, and immunodeficiency has been linked to life-threatening herpesvirus infections. Therefore, it was hypothesized that marijuana smoking may compromise immunity against herpesviruses leading to frequent recurrent infections. However, specific studies examining the effect of cannabinoids on herpesvirus infections do not uniformly support this hypothesis. While some early studies linked marijuana use to more frequent reactivation of herpes simplex virus 2 (HSV-2) infections, most recent studies show that THC is a potent inhibitor of lytic replication of oncogenic gamma herpesviruses. The first part of this chapter will summarize current advances in cannabinoid research and herpesvirus biology and molecular genetics. The second section will summarize in detail the literature dealing specifically with interaction of cannabinoids with herpesviruses.

JERRY L. BULEN and PETER G. MEDVECZKY • Department of Medical Microbiology and Immunology, University of South Florida College of Medicine, Tampa, FL 33612-4799.

Infectious Diseases and Substance Abuse, edited by Herman Friedman *et al.*
Springer, New York, 2005.

2. CANNABINOID LIGANDS AND RECEPTORS

The mechanism, by which cannabinoids produce their broad array of physiological effects was initially thought to be caused by nonspecific interactions of the highly lipophilic cannabinoid compounds and the lipid bilayer of the cell membrane thus causing a disruption of membrane processes.[3] This view was modified by identification of cannabinoid receptors. The finding of negative regulation of adenylate cyclase following exposure to cannabinoids, an enzyme normally associated with membrane-bound receptors in mammalian systems,[4,5] was instrumental in the cloning of the first cannabinoid receptor (CB1) in 1990 from a rat brain cDNA library.[6] This cloned cDNA encodes a 473 amino acid protein with the features of a G-protein-coupled receptor. A second major form of the receptor (cannabinoid receptor type 2 [CB2]) has been isolated and cloned from the promyelocytic line HL60.[7] The two receptors have an approximate 68% identity within their transmembrane domain, the portion of the receptor involved in ligand binding.

The primary psychoactive component of marijuana smoke is THC, which was historically the compound used in the early experiments. Synthetic agonists have since been developed that have varied characteristics with either specific binding for a single cannabinoid receptor (CB1 or CB2) or both. These ligands are classified as classical cannabinoids or Δ^9-THC-like compounds, nonclassical cannabinoids, and aminoalkylindoles. Several endogenous ligands for cannabinoid receptors have also been identified, most notably arachidonoylethanolamide (anandamide); these are grouped together in the eicosanoid group of cannabinoid receptor agonists.[8] Cannabinoid antagonists have also been developed with the prototypic members of this class of this series of compounds SR141716A, a potent CB1-selective ligand, and SR144528, a potent CB2-selective ligand (Table I). These compounds prevent or reverse the effects mediated by the CB1 and CB2 receptors.[9,10]

CB1 is the primary type of cannabinoid receptor found in the CNS but is only modestly expressed in the immune system.[24–26] The CB2 receptor appears to be the predominant form of cannabinoid receptor of the immune system and is conspicuously absent from the CNS.[7,26] The distribution pattern of CB2 mRNA in the human blood cell population has been determined, with a rank order of B lymphocytes > Natural killer cells > > monocytes > PMNs > T8 lymphocytes > T4 lymphocytes.[27]

3. REGULATION OF ADENYLATE CYCLASE BY CANNABINOID RECEPTORS

Modulation of adenylate cyclase by cannabinoids has been demonstrated in virtually every cell-line and tissue that expresses functional cannabinoid receptors as well as cell-lines initially devoid of either CB1 or CB2 and then successfully transfected with either of the two receptor genes.[6,28,29] The use of the human T-cell line, Jurkat E6-1, which has been shown to express nonfunctional

TABLE I
List of Cannabinoid Ligands Ordered by
their Relative Selectivity

Ligand	References
CB1-selective ligands in order of	
decreasing CB1/CB2 selectivity	
ACEA	[11]
O-1812	[12]
SR141716A	[13]
AM281	[14]
ACPA	[11]
2-Arachidonylglyceryl ether	[15]
LY320135	[13]
R-(+)-methanandamide	[16]
Nonspecific CB1/CB2 ligands	
Anandamide	[16]
	[17]
2-Arachidonoylglycerol	[18]
HU-201	[19]
CP55940	[20]
Δ^9-THC	[19]
Δ^8-THC	[21]
R-(+)-Win55212	[17]
CB2-selective ligands in order of	
increasing CB2/CB1 selectivity	
JWH-015	[17]
JWH-051	[22]
AM 630	[20]
L-759656	[20]
HU-308	[23]
SR144528	[9]

CB2 receptors and no CB1 receptors,[26] is resistant to modulation of adenylate cyclase by cannabinoids. These studies with Jurkat E6-1 cells helped establish that modulation of adenylate cyclase activity by cannabinoids is not mediated by non-specific membrane actions but are dependant on functional receptor activity.

CB1 and CB2 negatively regulate adenylate cyclase through a pertussis toxin sensitive G-protein-coupled receptor.[21] The cAMP cascade is regulated by the formation of cAMP from ATP by adenylate cyclase. The cAMP then binds to the regulatory subunits of protein kinase A (PKA), which results in the release and activation of PKA-catalytic subunits. These catalytic subunits go on to phosphory-late a variety of intracellular proteins including the cAMP response element bind-ing protein/activating transcription factor (CREB/ATF). CREB is activated by PKA-mediated phosphorylation and forms either homo- or heterodimers with a variety of other transcription factors. Several laboratories have shown that both the Fos and Jun family members can dimerize with CREB, and that these het-erodimers are capable of binding AP-1 sites.[30,31] PKA has also been implicated in the activation of the NF-κB/Rel family of transcription factor (Fig. 1).

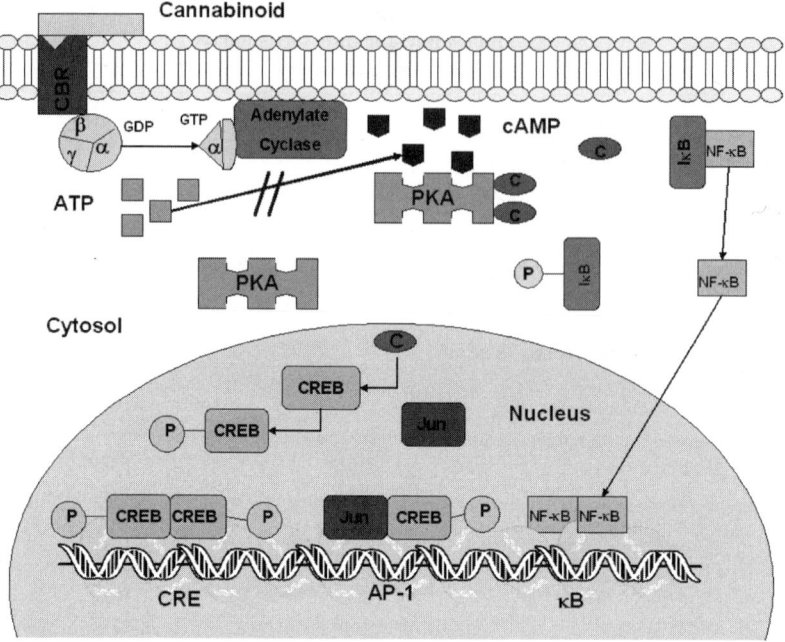

FIGURE 1. Proposed model of the cannabinoid receptor and its modulation of certain transcription factors by inhibition of cyclic AMP.

4. HERPESVIRUSES

Herpesviridae are a group of animal viruses that are ubiquitous to the vertebrate species. They are characterized as large, enveloped double-stranded (ds) DNA viruses with genomes in the range of 120–250 kb.[32] Herpesvirus infections of humans are a major public health problem due to their prevalence in the population. There are eight known human herpesviruses (HHVs) that have been identified (HHV 1–8) and most of them produce primary infections that are asymptomatic, leading to widespread transmission early in life. Notable exceptions to this asymptomatic trend are chicken pox (caused by Varicella-zoster virus [VZV]), mononucleosis (caused by Epstein–Barr virus [EBV]), and genital or neonatal, disseminated herpes (caused by HSV-2). In most other cases, the major episodes of symptomatic disease are secondary to viral reactivation from latency, including shingles following reactivation of latent VZV and "cold sores" with HSV-1 reactivation. Reactivation of herpesviruses from latency can also produce some more severe conditions including HSV-1 encephalitis and keritoconjunctivitis, which causes blindness. Immune compromised patients including organ transplant recipients or AIDS patients are particularly at high risk for severe infection secondary to herpesvirus reactivation. Notably, EBV and Kaposi's sarcoma-associated herpesvirus (KSHV) have been implicated as the causative agents of several diverse malignancies of leukamoid or epithelial origin, also associated with immunocompromised patients.

Members of the *Herpesviridae* family of viruses were initially classified into three subfamilies based on their biologic properties, *alpha-, beta-,* and *gammaherpesvirinae*. The subfamily *alphaherpesvirinae* is characterized by a variable host range, relatively short reproductive cycle, and rapid spread in culture, efficient destruction of infected cells, and the capacity to establish latency primarily, but not exclusively, in sensory ganglia. Examples of human α-herpesviruses include HSV-1, HSV-2, and VZV. The characteristics of *betaherpesvirinae* include a restricted host range, a long reproductive cycle that progresses slowly in culture, and infected cells that frequently become enlarged (cytomegalia). These viruses can maintain latency in secretory glands, lymphoreticular cells, kidneys, and other tissues. Human examples include *Cytomegalovirus* (CMV), HHV-6, and HHV-7. The *gammaherpesvirinae* have a very limited host range, with the experimental host range being limited to the family or order of the natural host. All of these viruses, *in vitro*, replicate in lymphoblastoid cells, and some cause lytic infections in specific types of epithelial and fibroblastic cells. These viruses are even specific for B- or T lymphocytes. The γ-herpesviruses contain two genera each containing one virus that has been found to be human pathogens, *Lymphocryptovirus* (EBV) and *Rhadinovirus* (KSHV).

4.1. α-Herpesvirus Latency and Reactivation

Following the initial infection of rodents, rabbits, or humans with HSV-1, productive infection is initiated in the mucosal epithelium. Virus particles or subparticles that are released then enter sensory neurons and are transported to the sensory ganglia. The trigeminal ganglia (TG) are the primary sites for latency since most of the initial inoculations of HSV-1 occur at oral, nasal, or ocular sites.[33] The establishment of latency in neurons is essentially a passive phenomenon where no viral gene product is involved in the process. The failure of the productive cycle in sensory neurons has been related to altered or absent specific transcription factors found in neuronal cells compared to non-neuronal cells, notably by interfering with the first activation via VP-16.[34–37] Viral mutants lacking functional ICP4 or VP-16 genes, which are unable to express few if any productive cycle proteins, have been shown to establish latent infections.[38–43]

During latency, the only abundant viral RNAs produced are the latency-associated transcripts (LATs).[44] In contrast to the other HSV promoters, the promoter that directs the expression of LAT is activated in sensory neurons. There are two separate promoter fragments that are located upstream of the start site of LAT, latency-associated promoters 1 and 2 (LAP 1 & 2).[45,46] LAP1 is critical for directing expression in sensory neurons,[46–48] and LAP2 promotes expression of the stable 2-kb LAT during the productive phase of HSV-1 in cultured cells.[46,49] The LAT promoter has an abundance of cellular transcription factor binding sites, and exhibits both neuronal and nonneuronal specific expression of a reporter gene in transient-transfection assays.[50,51] The two cAMP-responsive elements (CRE) binding sites in the LAT promoter are functional because it has been shown that cAMP activates the promoter.[52,53] The CRE element that is proximal to the TATA box is important for LAT expression in neurons, and its presence has a positive effect on reactivation from latency.[52,54,55]

The various functions of LATs have been described and each of these putative functions has been disputed.[56–65]

1. LATs block productive gene expression by blocking the transcription of the gene encoding ICP0 located in an antisense position from the stable 2.0 kb LAT or by other mechanisms.[56,57]
2. LATs enable reactivation from a latent state.[66]
3. LATs maintain the virus in a latent state.[67–69]

The LATs also encode a number of putative ORFs, and these encoded proteins may trigger one or more of the functions of the LATs. This mechanism has largely been discounted in the past, because mutations within the LAT ORFs do not affect the latency phenotype in animal models,[70,71] and no LAT-encoded proteins have been reliably detected in latently infected neurons.[72,73] It has recently been shown that LAT ORF expression overcomes cell-mediated repression of exogenous promoters in the HSV genome using a mechanism similar to that of ICP0.[66] Another function of LATs that has been shown recently is that they protect neurons from apoptosis.[67,74,75] A recent paper demonstrated that neuronal transcription factors can regulate the ICP0 promoter *in vivo*, in the absence of other viral proteins, either VP-16 or LATs, and that this property can be altered by changes in the physiological environment of the same neurons.[76] ICP0 is a key factor in the productive cycle of HSV because it has the ability to activate the expression of all classes of viral genes. The amino terminus of ICP0 is required for IE promoter activation, and a separate domain activates E or L promoters.[77,78] ICP0 also binds several cellular proteins including cyclin D3,[79] ubiquitin-specific protease[80,81] and elongation factor.[82] The protein binding activities of ICP0 have been shown to promote virus replication in differentiated cells.[83]

4.2. γ-Herpesvirus Latency and Reactivation

Lymphocytes latently infected or immortalized with a gammaherpesvirus (KSHV, EBV, herpesvirus saimiri [HVS], and murine herpesvirus-68 [MHV-68]) carry multiple copies of the viral genome as an episome, and can be propagated in culture for years without significant lytic viral production.[84–86] Studies have shown that viral episome replication in these latently infected dividing cells is mediated by host DNA polymerase.[85,86] This was demonstrated using antiviral drugs that inhibit thymidine kinase and viral DNA polymerase. These enzymes are expressed and utilized by KSHV only during productive, lytic infection.[87] After prolonged treatment with these drugs, TPA-induced linear KSHV DNA production was inhibited while the episomal or latent form of the viral genome was not affected.[85]

Spontaneous reactivation and production of virus has been demonstrated in a small number of some but not all immortalized cell lines.[88] The majority of the KSHV genome remains silent during latency. This is likely due to methylation of promoter sequences.[89] KSHV Rta (also known as Lyta), encoded by ORF 50, is necessary[90] and sufficient[91] to activate the lytic cycle. Rta is able to activate its own promoter[91,92] giving a rapid autocatalytic rise in expression. Cellular factors are also likely to be important for the reactivation of KSHV from latency. CREB binding protein (CBP) and c-Jun bind to Rta and activate Rta-mediated transcription[93] while Rta activation of the KSHV thymidine kinase promoter is dependant on Sp1.[94] The promoter of ORF 50 is heavily methylated in latent PEL cell lines and infected peripheral blood mononuclear cells.[89] Demethylation of the promoter is induced by TPA treatment explaining how this agent activates the lytic cycle of KSHV.[89] It is still unknown what the triggers of demethylation might be *in vivo*, but they are likely to be an indicator of cellular stress.

5. CANNABINOIDS AND HERPES SIMPLEX VIRUSES

The ability of marijuana to alter the course of viral disease was first reported by Juel-Jensen in 1972. This work associated an increase of HSV recurrence rates with marijuana smoking.[95] The study contained a small subject pool and relied on patient histories, but it opened the door for examinations of the relationship between cannabinoids, immunity, and the progression of viral infection.

Shortly after this initial work, a series of experimental studies using animal models of HSV infection to examine the effects of cannabinoids on viral progression was published. Morahan reported that THC decreased the resistance of mice to HSV following drug exposure.[96] Later, studies by Cabral showed a similar decrease in resistance in guinea pigs to vaginal infection by HSV-2 following THC exposure.[97] Cabral *et al.* followed this work with studies demonstrating that THC diminished the production of interferon α/β in mice during HSV infection, decreased the cytotoxic T-lymphocyte response to HSV, and caused a diminished macrophage "extrinsic" (not interferon mediated) antiviral activity.[98–100]

It has also been reported that high, "nonphysiologic" concentrations of THC inhibited the replication of HSV *in vitro* using direct exposure of the virus to the cannabinoid[101,102] at 50–100 µg/ml; the highest blood levels measured in marijuana smokers reach only perhaps 1 µg/ml. This effect was postulated to be due to nonspecific interactions of the lipophilic cannabinoid with the cellular membrane or the viral envelope.

6. THC INHIBITS KSHV AND EBV LYTIC REPLICATION

The most recent studies investigated the effects of THC, at "physiologic" concentrations, on gammaherpesvirus replication and reactivation. Because this drug has been shown to modulate various biochemical functions of lymphocytes,[1,2] the effect of THC on KSHV- and EBV-transformed lymphocytes was investigated.[103]

The BCBL-1 cell line, which spontaneously produces small amounts of KSHV, is suitable to determine if drugs induce or inhibit virus replication. BCBL-1 cells were grown in media with or without THC at varying concentrations, while control cells were grown in DMSO solvent alone. Following an incubation period of 48 hr, the cells were analyzed using the Gardella method[91,104] that efficiently separates the slowly migrating episomal DNA (latent genome), and the rapid migrating linear DNA (actively replicating genome). THC showed inhibition of linear but not episomal KSHV DNA in BCBL-1 cells.[103] The BC-3 cell line was also tested and THC was shown to have a similar inhibitory effect on KSHV spontaneous reactivation. The 50% inhibitory concentration (Ki) of THC on BCBL-1 and BC-3 cells was calculated as 1 and 2.5 μg/ml, respectively.

Similar experiments as the one described for KSHV were performed with the transformed cell line P3HR1.[103] It was necessary to use the phorbol ester, TPA, to induce reactivation of the EBV in this cell line because they do not reactivate spontaneously. Cultures of P3HR1 were also grown with various concentrations of THC or DMSO control and TPA was added to stimulate EBV reactivation. After an incubation period, the cells were analyzed using the Gardella method as previously mentioned. The results showed that THC blocked the TPA-induced reactivation of EBV in P3HR1 cells (Ki around 1 μg/ml) while it had no effect on the episomal viral genome.

6.1. THC Inhibits MHV-68 and HVS Lytic Replication in Monolayer Cells

Surprisingly, THC also inhibits virus production of gammaherpesviruses in non-lymphoid cells.[103] THC strongly inhibits both the cytopathic lytic effects of MHV-68 and HVS in NIH 312 monolayer cells. NIH 312 cells were infected with MHV-68 in the presence of various concentrations of THC (in DMSO), while control cultures were treated with the DMSO solvent alone. Additional controls included uninfected cells grown in the presence of THC or DMSO. Cell cultures were then incubated for 48 hr and then examined microscopically for cytopathic effects. The full cytopathic effect of MHV-68 was seen in the control infected cell culture treated with the DMSO solvent alone. A majority of the adherent cells were detached from the plate and the remaining, loosely adherent cells, showed morphological changes consistent with cytopathic effects including increased density and loss of the normal spindle shape (rounding) when compared to the uninfected controls. Infected cells cultured in the presence of THC at concentrations from 1.25 to 10 μg/ml were indistinguishable from uninfected controls. The effect of 0.6 μg/ml of THC was found to be intermediate between uninfected cells and the full cytopathic effect seen with infected cells cultured in the presence of the DMSO solvent alone. Similar results were obtained using HVS in Owl Monkey Kidney (OMK) cells. Yield reduction assays were performed to quantitatively determine the antiviral effects of THC.[103] Virus yield was significantly suppressed by THC with a suppression of over 200-fold at 10 μg/ml of THC. The 50% inhibitory concentration was estimated at 0.6 μg/ml. Similar results were obtained in two separate experiments and with HVS in OMK cells.

6.2. THC is Not Cytotoxic to Murine NIH 312 or OMK Cells and Does Not Inhibit HSV-1 Lytic Replication in Monolayer Cells

To rule out nonspecific cytotoxic effects as the source of the decreased virus yield observed with THC treatment, Medveczky et al.[103] also tested whether THC altered the cell division or morphology of NIH 312 and OMK cells. Monolayers of these cells were prepared at and cultured in the presence of THC at concentrations ranging from 0.6 to 10 μg/ml. The THC-treated cultures were indistinguishable from control cultures. They formed confluent monolayers and showed no evidence of altered morphology. This showed that the observed antiviral effect of THC on gammaherpesvirus lytic replication was not due to cytotoxicity.

The observed antiviral effects of THC on gammaherpesviruses may also be nonspecific and may also inhibit lytic replication of HSV-1 in NIH 312 cells. To test this, the possible effect of THC was measured on the production of HSV-1.[103] IH 312 cells were infected with HVS-1, with or without THC dissolved in DMSO. Control cultures were treated with DMSO solvent alone. All cultures were incubated for 24 hr and then the virus was harvested and titrated. THC had no inhibitory effect on replication of HSV-1 in NIH 312 or OMK cells. Therefore, THC specifically targets a viral or cellular component uniquely required for gammaherpesvirus lytic replication.

6.3. THC Inhibits ORF 50 mRNA Transcription Initiation

As previously mentioned, the ORF 50 gene product of KSHV and HVS is necessary[90] and sufficient to activate the lytic cycle of KSHV and HVS replication. Unpublished experimental data show that the KSHV ORF 50 mRNA transcription decreased in the presence of THC compared to control while the drug did not effect actin expression (M. Medveczky, T. W. Klein, and P. G. Medveczky, unpublished observations). Furthermore, THC specifically inhibited the ORF 50 promoter activity of KSHV and MHV-68 as tested in luciferase reporter assays while the CMV immediate early promoter activity was not affected by the drug (M. Medveczky, T. W. Klein, and P. G. Medveczky, unpublished observations). Therefore, THC reduces ORF 50 transcription initiation.

7. SPLEEN CELLS FROM CB1 KNOCKOUT MICE ARE MORE SUSCEPTIBLE TO LYTIC MHV-68 INFECTION THAN CELLS FROM WILD-TYPE MICE, AND KNOCKOUT SPLENOCYTES DO NOT SUPPORT GENERATION OF CIRCULAR (LATENT) EPISOMES

It was hypothesized that cannabinoid receptor knockout animals may have an increased susceptibility to gammaherpesvirus infection if either one of these receptors are required for inhibition of lytic infection by endocannabinoids produced by most cells (M. Medveczky, T. W. Klein, and P. G. Medveczky, unpublished observations). Utilizing spleen cells harvested from wild-type, hetero-, and homozygous CB1 knockout mice infected with MHV-68, it was attempted to determine what effect, if any, the lack of the CB1 receptor would have on virus replication. CB1 knockout or wild-type spleen cells were infected with MHV-68 and harvested at intervals of 1, 3, and 6 days postinfection, then subject to Gardella gel and Southern blot analysis. Results of the Southern blot showed that the amount of linear viral DNA indicating lytic infection gradually decreased over time in the wild-type CB1 samples while the amount of linear viral DNA increased with time in the homozygous CB1 knockout mouse samples. The blot also revealed latent episomal viral DNA in wild-type spleen cells but episomal DNA was absent in splenocytes from either the hetero- or homozygous animals. Splenocytes from knockout animals died as a result of lytic viral infection by day 6 after infection.

In conclusion, based on the above experiment, CB1 receptors appear to play a role in the establishment of the episomal, latent genomes of MHV-68 since episomal DNA was only detected in splenocytes from wild-type mice. It can also be inferred that the CB1 receptors also provide defense against lytic killing by the MHV-68 virus as CB1 knockout cells died sooner than cells from wild-type mice. These data suggest that the cannabinoid system is somehow involved in regulating lytic and latent gammaherpesvirus infections. It is possible that splenocytes constitutively, or in response to herpesvirus infection, produce endocannabinoids. These compounds may bind the CB receptors and cause downregulation of lytic replication. Substantial future experimentation is required to evaluate if this hypothesis is correct.

8. CONCLUSIONS

During the past decade, much progress has been made in the field of molecular biology and functional analysis regarding cannabinoid receptors as well as the viral and cellular mechanisms that control herpesvirus replication, latency, and reactivation. THC has been shown to be nonselective to cannabinoid receptors. Several synthetic, potent, agonists have since been discovered which can be selective, stimulating either CB1 or CB2 receptors, or nonselective agonists, stimulating both CB1 and CB2 receptors.

Early studies have attempted to evaluate if THC has effect on alphaherpesvirus replication. Since there is no clear conclusion if THC and cannabinoids inhibit or enhance replication/reactivation of these viruses, investigation should continue in light of new advances in herpesvirus molecular biology and cannabinoid research.

In light of earlier studies with HSV-1 and HSV-2, it is surprising that THC is a quite potent and selective inhibitor of gammaherpesvirus replication. Table II shows that THC is more potent and often more selective inhibitor of KSHV and MHV-68 than several licensed antiviral drugs including ganciclovir, acyclovir, and foscarnet. It must be stressed, however that these findings are based on *in vitro* experiments and antiviral effects of THC have not been confirmed in experimental animals.

TABLE II
Comparison of 50% Antiviral and Cell Division Inhibitory Concentrations (IC$_{50}$)
and Selectivity (Viral vs Cellular IC$_{50}$) of Selected Antiviral Drugs and THC

Inhibitor compound	KSHV IC$_{50}$(μM)	Cellular IC$_{50}$(μM)	Selectivity index (viral vs cellular IC$_{50}$)
Acyclovir	75 (ref. 85)	Not done	
Ganciclovir	5.1 (ref. 85)	Not done	
Foscarnet	97 (ref. 85)	Not done	
Cidofovir	0.05 (ref. 85)	Not done	
THC	3 (ref. 103)	30 (ref. 103)	10 (ref. 103)
	MHV-68 IC$_{50}$(μM)		
Acyclovir	6 (ref. 105)	182 (ref. 105)	30 (ref. 105)
	100 (ref. 106)		
Ganciclovir	28 (ref. 105)	108 (ref. 105)	3.7 (ref. 105)
Foscarnet	120 (ref. 105)	1413 (ref. 105)	11 (ref. 105)
Cidofovir	0.08 (ref. 105)	78 (ref. 105)	10,000 (ref. 105)
THC	1.9 (ref. 103)	90 (ref. 103)	47 (ref. 103)

Interestingly, statistical analysis indicates lower incidence of Kaposi's sarcoma in HIV-positive women using nonintravenous drugs.[107] About 5.4% of HIV-positive women with no drug use developed KS whereas none of the 47 women in this study who used only marijuana suffered from KS[107] James Goedert, personal communication). This report, however, involved relatively few individuals so further analysis of a larger cohort is warranted.

As outlined in this chapter, recent findings indicate a connection between gammaherpesvirus replication, THC, and the cannabinoid system. THC inhibits lytic replication and reactivation of gammaherpesviruses, and the cannabinoid system has a generally negative controlling effect. The mechanism by which THC exerts its effect appears to involve signaling through the cannabinoid receptors; however, the exact mechanism of this inhibition requires further investigation. One mechanism may involve the interaction of the ORF 50 gene product (Rta) and the CBP. As mentioned previously, KSHV Rta has been shown to activate its own promoter,[91,92] and the CREB binding protein has been shown to bind to Rta and activate Rta-mediated transcription.[93] Cannabinoid receptor binding has been shown to downregulate the level of activated CREB through a decrease of cyclic AMP,[26] therefore, inhibiting Rta-mediated transcription. Transcription of ORF 50 mRNA is decreased in the presence of THC when compared to controls. The exact mechanism of this inhibition needs also to be further investigated. Is the inhibition at the level of protein binding, translation, mRNA transcription, DNA methylation, or chromatin stabilization are major questions that remain to be answered. Also, THC may inhibit other genes that are necessary for gammaherpesvirus lytic replication.

We believe that studies on cannabinoids and herpesviruses are worth continuing because there are obvious potential benefits. Better understanding may lead to the development of specific nonpsychoactive drugs that may inhibit herpesvirus reactivation or even possibly eradicate the virus completely by preventing the establishment of latency.

Data indicating that THC inhibits MHV-68 replication and the virus cannot establish latent genomes in CB1 knockout mice are also significant. Further studies to evaluate the effect of THC and cannabinoid receptor stimulation have on MHV-68 infection and latency in murine models. This should include experiments to evaluate if and how cannabinoids modulate various aspects of MHV-68 infection and disease progression *in vivo*, and whether cannabinoid receptor activation, particularly by endocannabinoid stimulation, plays a role in controlling latent infections. Specific questions that remain to be answered include: Can cannabinoids prevent and suppress MHV-68 infection? Can cannabinoids influence an established MHV-68 infection? Can cannabinoids cause immunosuppression competing with potential antiviral effects causing a variance in the course of MHV-68 infection? What is the course of MHV-68 infection in cannabinoid receptor knockout mice?

Obviously the murine system can provide not only valuable insights on how THC interacts with herpesvirus infection, the cannabinoid, and the immune system, but could also be developed to test antiviral drugs exploiting the cannabinoid system. Ultimately, future studies could play a role in the development of novel nonpsychoactive antiviral drugs targeting the cannabinoid system.

REFERENCES

1. Klein, T., Friedman, H., and Specter, S. (1998). Marijuana, immunity and infection. *J. Neuroimmunol.* **83**:102–115.
2. Klein, T., Newton, C., and Friedman, H. (1998). Cannabinoid receptors and immunity. *Immunol. Today* **19**:373–381.
3. Kaminski, N. E. (1998). Regulation of the cAMP cascade, gene expression and immune function by cannabinoid receptors. *J. Neuroimmunol.* **83**(1–2):124–132.
4. Howlett, A. C. and Fleming, R. M. (1984). Cannabinoid inhibition of adenylate cyclase. Pharmacology of the response in neuroblastoma cell membranes. *Mol. Pharmacol.* **26**(3):532–538.
5. Howlett, A. C. (1985). Cannabinoid inhibition of adenylate cyclase. Biochemistry of the response in neuroblastoma cell membranes. *Mol. Pharmacol.* **27**(4):429–436.
6. Matsuda, L. A. *et al.* (1990). Structure of a cannabinoid receptor and functional expression of the cloned cDNA. *Nature* **346**(6284):561–564.
7. Munro, S., Thomas, K. L., and Abu-Shaar, M. (1993). Molecular characterization of a peripheral receptor for cannabinoids. *Nature* **365**(6441):61–65.
8. Devane, W. A. *et al.* (1992). Isolation and structure of a brain constituent that binds to the cannabinoid receptor. *Science* **258**(5090):1946–1949.
9. Rinaldi-Carmona, M. *et al.* (1998). SR 144528, the first potent and selective antagonist of the CB2 cannabinoid receptor. *J. Pharmacol. Exp. Ther.* **284**(2):644–650.
10. Rinaldi-Carmona, M. *et al.* (1994). SR141716A, a potent and selective antagonist of the brain cannabinoid receptor. *FEBS Lett.* **350**(2–3):240–244.
11. Hillard, C. J. *et al.* (1999). Synthesis and characterization of potent and selective agonists of the neuronal cannabinoid receptor (CB1). *J. Pharmacol. Exp. Ther.* **289**(3):1427–1433.
12. Di Marzo, V. *et al.* (2001). Highly selective CB(1) cannabinoid receptor ligands and novel CB(1)/VR(1) vanilloid receptor "hybrid" ligands. *Biochem. Biophys. Res. Commun.* **281**(2):444–451.
13. Felder, C. C. and Glass, M. (1998). Cannabinoid receptors and their endogenous agonists. *Annu. Rev. Pharmacol. Toxicol.* **38**:179–200.

14. Lan, R. *et al.* (1999). Design and synthesis of the CB1 selective cannabinoid antagonist AM281: A potential human SPECT ligand. *AAPS PharmSci.* **1**(2).

15. Hanus, L. *et al.* (2001). 2-Arachidonyl glyceryl ether, an endogenous agonist of the cannabinoid CB1 receptor. *Proc. Natl. Acad. Sci. USA.* **98**(7):3662–3665.

16. Lin, S. *et al.* (1998). Novel analogues of arachidonylethanolamide (anandamide): Affinities for the CB1 and CB2 cannabinoid receptors and metabolic stability. *J. Med. Chem.* **41**(27):5353–5361.

17. Showalter, V. M. *et al.* (1996). Evaluation of binding in a transfected cell line expressing a peripheral cannabinoid receptor (CB2): Identification of cannabinoid receptor subtype selective ligands. *J. Pharmacol. Exp. Ther.* **278**(3):989–999.

18. Mechoulam, R. *et al.* (1995). Identification of an endogenous 2-monoglyceride, present in canine gut, that binds to cannabinoid receptors. *Biochem. Pharmacol.* **50**(1):83–90.

19. Felder, C. C. *et al.* (1995). Comparison of the pharmacology and signal transduction of the human cannabinoid CB1 and CB2 receptors. *Mol. Pharmacol.* **48**(3):443–450.

20. Ross, R. A. *et al.* (1999). Agonist-inverse agonist characterization at CB1 and CB2 cannabinoid receptors of L759633, L759656, and AM630. *Br. J. Pharmacol.* **126**(3):665–672.

21. Busch-Petersen, J. *et al.* (1996). Unsaturated side chain beta-11-hydroxyhexahydrocannabinol analogs. *J. Med. Chem.* **39**(19):3790–3796.

22. Huffman, J. W. *et al.* (1996). Synthesis and pharmacology of a very potent cannabinoid lacking a phenolic hydroxyl with high affinity for the CB2 receptor. *J. Med. Chem.* **39**(20):3875–3877.

23. Hanus, L. *et al.* (1999). HU-308: A specific agonist for CB(2), a peripheral cannabinoid receptor. *Proc. Natl. Acad. Sci. USA* **96**(25):14228–14233.

24. Bouaboula, M. *et al.* (1993). Cannabinoid-receptor expression in human leukocytes. *Eur. J. Biochem.* **214**(1):173–180.

25. Kaminski, N. E. *et al.* (1992). Identification of a functionally relevant cannabinoid receptor on mouse spleen cells that is involved in cannabinoid-mediated immune modulation. *Mol. Pharmacol.* **42**(5):736–742.

26. Schatz, A. R. *et al.* (1997). Cannabinoid receptors CB1 and CB2: A characterization of expression and adenylate cyclase modulation within the immune system. *Toxicol. Appl. Pharmacol.* **142**(2):278–287.

27. Galiegue, S. *et al.* (1995). Expression of central and peripheral cannabinoid receptors in human immune tissues and leukocyte subpopulations. *Eur. J. Biochem.* **232**(1):54–61.

28. Slipetz, D. M. *et al.* (1995). Activation of the human peripheral cannabinoid receptor results in inhibition of adenylyl cyclase. *Mol. Pharmacol.* **48**(2):352–361.

29. Bayewitch, M. *et al.* (1995). The peripheral cannabinoid receptor: Adenylate cyclase inhibition and G protein coupling. *FEBS Lett.* **375**(1–2):143–147.

30. Ivashkiv, L. B. *et al.* (1990). mXBP/CRE-BP2 and c-Jun form a complex which binds to the cyclic AMP, but not to the 12-O-tetradecanoylphorbol-13-acetate, response element. *Mol. Cell. Biol.* **10**(4):1609–1621.

31. Hai, T. and Curran, T. (1991). Cross-family dimerization of transcription factors Fos/Jun and ATF/CREB alters DNA binding specificity. *Proc. Natl. Acad. Sci. USA* **88**(9):3720–3724.

32. Roizman, B. and Knipe, D. M. (2001). Herpes simplex viruses and their replication. In B. N. Fields *et al.* (eds), *Fields Virology*, Lippincott Williams & Wilkins, Philadelphia, 2399–2459.

33. Baringer, J. R. and Swoveland, P. (1973). Recovery of herpes-simplex virus from human trigeminal ganglions. *N. Engl. J. Med.* **288**(13):648–650.

34. Wheatley, S. C. *et al.* (1992). Elevation of cyclic AMP levels in cell lines derived from latently infectable sensory neurons increases their permissivity for herpes virus infection by activating the viral immediate-early 1 gene promoter. *Brain Res. Mol. Brain Res.* **12**(1–3):149–154.

35. Lillycrop, K. A. *et al.* (1994). Inhibition of herpes simplex virus infection by ectopic expression of neuronal splice variants of the Oct-2 transcription factor. *Nucleic Acids Res.* **22**(5):815–820.

36. Lillycrop, K. A. *et al.* (1991). The octamer-binding protein Oct-2 represses HSV immediate-early genes in cell lines derived from latently infectable sensory neurons. *Neuron* 7(3):381–390.

37. Howard, M. K. *et al.* (1993). Transactivation by the herpes simplex virus virion protein Vmw65 and viral permissivity in a neuronal cell line with reduced levels of the cellular transcription factor Oct-1. *Exp. Cell. Res.* 207(1):194–196.

38. Valyi-Nagy, T. *et al.* (1991). Investigation of herpes simplex virus type 1 (HSV-1) gene expression and DNA synthesis during the establishment of latent infection by an HSV-1 mutant, in1814, that does not replicate in mouse trigeminal ganglia. *J. Gen. Virol.* 72(Pt 3):641–649.

39. Steiner, I. *et al.* (1990). A herpes simplex virus type 1 mutant containing a nontransducing Vmw65 protein establishes latent infection in vivo in the absence of viral replication and reactivates efficiently from explanted trigeminal ganglia. *J. Virol.* 64(4):1630–1638.

40. Sedarati, F., Margolis, T. P., and Stevens, J. G. (1993). Latent infection can be established with drastically restricted transcription and replication of the HSV-1 genome. *Virology.* 192(2):687–691.

41. Katz, J. P., Bodin, E. T., and Coen, D. M. (1990). Quantitative polymerase chain reaction analysis of herpes simplex virus DNA in ganglia of mice infected with replication-incompetent mutants. *J. Virol.* 64(9):4288–4295.

42. Harris, R. A. and Preston, C. M. (1991). Establishment of latency in vitro by the herpes simplex virus type 1 mutant in1814. *J. Gen. Virol.* 72(Pt 4):907–913.

43. Dobson, A. T. *et al.* (1990). A latent, nonpathogenic HSV-1-derived vector stably expresses beta-galactosidase in mouse neurons. *Neuron.* 5(3):353–360.

44. Stevens, J. G. *et al.* (1987). RNA complementary to a herpesvirus alpha gene mRNA is prominent in latently infected neurons. *Science.* 235(4792):1056–1059.

45. Goins, W. F. *et al.* (1994). A novel latency-active promoter is contained within the herpes simplex virus type 1 UL flanking repeats. *J. Virol.* 68(4):2239–2252.

46. Chen, X. *et al.* (1995). Two herpes simplex virus type 1 latency-active promoters differ in their contributions to latency-associated transcript expression during lytic and latent infections. *J. Virol.* 69(12):7899–7908.

47. Mitchell, W. J. *et al.* (1990). A herpes simplex virus type 1 variant, deleted in the promoter region of the latency-associated transcripts, does not produce any detectable minor RNA species during latency in the mouse trigeminal ganglion. *J. Gen. Virol.* 71(Pt 4):953–957.

48. Dobson, A. T. *et al.* (1989). Identification of the latency-associated transcript promoter by expression of rabbit beta-globin mRNA in mouse sensory nerve ganglia latently infected with a recombinant herpes simplex virus. *J. Virol.* 63(9):3844–3851.

49. Nicosia, M. *et al.* (1994). The HSV-1 2-kb latency-associated transcript is found in the cytoplasm comigrating with ribosomal subunits during productive infection. *Virology* 204(2):717–728.

50. Zwaagstra, J. C. *et al.* (1990). Activity of herpes simplex virus type 1 latency-associated transcript (LAT) promoter in neuron-derived cells: Evidence for neuron specificity and for a large LAT transcript. *J. Virol.* 64(10):5019–5028.

51. Batchelor, A. H. and O'Hare, P. (1990). Regulation and cell-type-specific activity of a promoter located upstream of the latency-associated transcript of herpes simplex virus type 1. *J. Virol.* 64(7):3269–3279.

52. Leib, D. A. *et al.* (1991). The promoter of the latency-associated transcripts of herpes simplex virus type 1 contains a functional cAMP-response element: Role of the latency-associated transcripts and cAMP in reactivation of viral latency. *Proc. Natl. Acad. Sci. USA* 88(1):48–52.

53. Kenny, J. J. *et al.* (1994). Identification of a second ATF/CREB-like element in the herpes simplex virus type 1 (HSV-1) latency-associated transcript (LAT) promoter. *Virology* 200(1):220–235.

54. Rader, K. A. *et al.* (1993). In vivo characterization of site-directed mutations in the promoter of the herpes simplex virus type 1 latency-associated transcripts. *J. Gen. Virol.* 74(Pt 9):1859–1869.

55. Bloom, D. C. *et al.* (1997). Mutagenesis of a cAMP response element within the latency-associated transcript promoter of HSV-1 reduces adrenergic reactivation. *Virology* **236**(1):202–207.

56. Garber, D. A., Schaffer, P. A., and Knipe, D. M. (1997). A LAT-associated function reduces productive-cycle gene expression during acute infection of murine sensory neurons with herpes simplex virus type 1. *J. Virol.* **71**(8):5885–5893.

57. Chen, S. H. *et al.* (1997). A viral function represses accumulation of transcripts from productive-cycle genes in mouse ganglia latently infected with herpes simplex virus. *J. Virol.* **71**(8):5878–5884.

58. Thompson, R. L. and Sawtell, N. M. (1997). The herpes simplex virus type 1 latency-associated transcript gene regulates the establishment of latency. *J. Virol.* **71**(7):5432–5440.

59. Trousdale, M. D. *et al.* (1991). In vivo and in vitro reactivation impairment of a herpes simplex virus type 1 latency-associated transcript variant in a rabbit eye model. *J. Virol.* **65**(12):6989–6993.

60. Leib, D. A. *et al.* (1989). Immediate-early regulatory gene mutants define different stages in the establishment and reactivation of herpes simplex virus latency. *J. Virol.* **63**(2):759–768.

61. Leib, D. A. *et al.* (1989). A deletion mutant of the latency-associated transcript of herpes simplex virus type 1 reactivates from the latent state with reduced frequency. *J. Virol.* **63**(7):2893–2900.

62. Hill, T. J., Blyth, W. A., and Harbour, D. A. (1982). Recurrent herpes simplex in mice; topical treatment with acyclovir cream. *Antiviral Res.* **2**(3):135–146.

63. Gordon, Y. J. *et al.* (1990). Host species and strain differences affect the ability of an HSV-1 ICP0 deletion mutant to establish latency and spontaneously reactivate in vivo. *Virology* **178**(2):469–477.

64. Clements, G. B. and Stow, N. D. (1989). A herpes simplex virus type 1 mutant containing a deletion within immediate early gene 1 is latency-competent in mice. *J. Gen. Virol.* **70**(Pt 9):2501–2506.

65. Block, T. M. *et al.* (1990). A herpes simplex virus type 1 latency-associated transcript mutant reactivates with normal kinetics from latent infection. *J. Virol.* **64**(7):3417–3426.

66. Thomas, S. K. *et al.* (2002). A protein encoded by the herpes simplex virus (HSV) type 1 2-kilobase latency-associated transcript is phosphorylated, localized to the nucleus, and overcomes the repression of expression from exogenous promoters when inserted into the quiescent HSV genome. *J. Virol.* **76**(8):4056–4067.

67. Thompson, R. L. and Sawtell, N. M. (2001). Herpes simplex virus type 1 latency-associated transcript gene promotes neuronal survival. *J. Virol.* **75**(14):6660–6675.

68. Sawtell, N. M. and Thompson, R. L. (1992). Herpes simplex virus type 1 latency-associated transcription unit promotes anatomical site-dependent establishment and reactivation from latency. *J. Virol.* **66**(4):2157–2169.

69. Perng, G. C. *et al.* (2000). The latency-associated transcript gene enhances establishment of herpes simplex virus type 1 latency in rabbits. *J. Virol.* **74**(4):1885–1891.

70. Farrell, M. J. *et al.* (1993). The herpes simplex virus type 1 reactivation function lies outside the latency-associated transcript open reading frame ORF-2. *J. Virol.* **67**(6):3653–3655.

71. Fareed, M. U. and Spivack, J. G. (1994). Two open reading frames (ORF1 and ORF2) within the 2.0-kilobase latency-associated transcript of herpes simplex virus type 1 are not essential for reactivation from latency. *J. Virol.* **68**(12):8071–8081.

72. Lagunoff, M. and Roizman, B. (1994). Expression of a herpes simplex virus 1 open reading frame antisense to the gamma(1)34.5 gene and transcribed by an RNA 3′ coterminal with the unspliced latency-associated transcript. *J. Virol.* **68**(9):6021–6028.

73. Doerig, C., Pizer, L. I., and Wilcox, C. L. (1991). An antigen encoded by the latency-associated transcript in neuronal cell cultures latently infected with herpes simplex virus type 1. *J. Virol.* **65**(5):2724–2727.

74. Perng, G. C. *et al.* (2000). Virus-induced neuronal apoptosis blocked by the herpes simplex virus latency-associated transcript. *Science.* **287**(5457):1500–1503.

75. Inman, M. *et al.* (2001). Region of herpes simplex virus type 1 latency-associated transcript sufficient for wild-type spontaneous reactivation promotes cell survival in tissue culture. *J. Virol.* **75**(8):3636–3646.

76. Loiacono, C. M., Myers, R. and Mitchell, W. J. (2002). Neurons differentially activate the herpes simplex virus type 1 immediate-early gene ICP0 and ICP27 promoters in transgenic mice. *J. Virol.* **76**(5):2449–2459.

77. Lium, E. K. and Silverstein, S. (1997). Mutational analysis of the herpes simplex virus type 1 ICP0 C3HC4 zinc ring finger reveals a requirement for ICP0 in the expression of the essential alpha27 gene. *J. Virol.* **71**(11):8602–8614.

78. Lium, E. K. *et al.* (1998). The NH_2 terminus of the herpes simplex virus type 1 regulatory protein ICP0 contains a promoter-specific transcription activation domain. *J. Virol.* **72**(10):7785–7795.

79. Kawaguchi, Y., Van Sant, C., and Roizman, B. (1997). Herpes simplex virus 1 alpha regulatory protein ICP0 interacts with and stabilizes the cell cycle regulator cyclin D3. *J. Virol.* **71**(10):7328–7336.

80. Meredith, M., Orr, A., and Everett, R. (1994). Herpes simplex virus type 1 immediate-early protein Vmw110 binds strongly and specifically to a 135-kDa cellular protein. *Virology* **200**(2):457–469.

81. Meredith, M. *et al.* (1995). Separation of sequence requirements for HSV-1 Vmw110 multimerisation and interaction with a 135-kDa cellular protein. *Virology* **209**(1):174–187.

82. Kawaguchi, Y., Bruni, R., and Roizman, B. (1997). Interaction of herpes simplex virus 1 alpha regulatory protein ICP0 with elongation factor 1delta:ICP0 affects translational machinery. *J. Virol.* **71**(2):1019–1024.

83. Cai, W. and Schaffer, P. A. (1991). A cellular function can enhance gene expression and plating efficiency of a mutant defective in the gene for ICP0, a transactivating protein of herpes simplex virus type 1. *J. Virol.* **65**(8):4078–4090.

84. Medveczky, P. (1995). Oncogenic transformation of T cells by herpesvirus saimiri. In *DNA Tumor Viruses: Oncogenic Mechanisms*, Plenum Press, New York, pp. 239–252.

85. Medveczky, M. M. *et al.* (1997). In vitro antiviral drug sensitivity of the Kaposi's sarcoma-associated herpesvirus. *AIDS* **11**(11):1327–1332.

86. Kieff, E. and Rickinson, A. B. (2001). Epstein–Barr virus and its replication, In B. N. Fields *et al.* (eds), *Fields Virology*, Lippincott Williams & Wilkins, Philadelphia, pp. 2511–2573.

87. Russo, J. J. *et al.* (1996). Nucleotide sequence of the Kaposi sarcoma-associated herpesvirus (HHV8). *Proc. Natl. Acad. Sci. USA* **93**(25):14862–14867.

88. Olsen, S. J. a. M. and Patrick, S. (1998). Kaposi's sarcoma-associated herpesvirus (KSHV/HHV8) and the etiology of KS. In *Herpesviruses and Immunity*, Plenum Press, New York, pp 115–147.

89. Chen, J. *et al.* (2001). Activation of latent Kaposi's sarcoma-associated herpesvirus by demethylation of the promoter of the lytic transactivator. *Proc. Natl. Acad. Sci. USA* **98**(7):4119–4124.

90. Lukac, D. M., Kirshner, J. R., and Ganem, D. (1999). Transcriptional activation by the product of open reading frame 50 of Kaposi's sarcoma-associated herpesvirus is required for lytic viral reactivation in B cells. *J. Virol.* **73**(11):9348–9361.

91. Gradoville, L. *et al.* (2000). Kaposi's sarcoma-associated herpesvirus open reading frame 50/Rta protein activates the entire viral lytic cycle in the HH-B2 primary effusion lymphoma cell line. *J. Virol.* **74**(13):6207–6212.

92. Deng, H., Young, A., and Sun, R. (2000). Auto-activation of the rta gene of human herpesvirus-8/Kaposi's sarcoma-associated herpesvirus. *J. Gen. Virol.* **81**(Pt 12): 3043–3048.

93. Gwack, Y. *et al.* (2001). CREB-binding protein and histone deacetylase regulate the transcriptional activity of Kaposi's sarcoma-associated herpesvirus open reading frame 50. *J. Virol.* **75**(4):1909–1917.

94. Zhang, L., Chiu, J., and Lin, J. C. (1998). Activation of human herpesvirus 8 (HHV-8) thymidine kinase (TK) TATAA-less promoter by HHV-8 ORF50 gene product is SP1 dependent. *DNA Cell Biol.* **17**(9):735–742.

95. Juel-Jensen, B. E. (1972). Cannabis and recurrent herpes simplex. *Br. Med. J.* **4**(835):296.

96. Morahan, P. S. *et al.* (1979). Effects of cannabinoids on host resistance to *Listeria monocytogenes* and herpes simplex virus. *Infect. Immun.* **23**(3):670–674.

97. Cabral, G. A. *et al.* (1986). Effect of delta 9-tetrahydrocannabinol on herpes simplex virus type 2 vaginal infection in the guinea pig. *Proc. Soc. Exp. Biol. Med.* **182**(2):181–186.

98. Cabral, G. A., Lockmuller, J. C., and Mishkin, E. M. (1986). Delta 9-tetrahydrocannabinol decreases alpha/beta interferon response to herpes simplex virus type 2 in the B6C3F1 mouse. *Proc. Soc. Exp. Biol. Med.* **181**(2):305–311.

99. Cabral, G. A. and Vasquez, R. (1992). Delta 9-tetrahydrocannabinol suppresses macrophage extrinsic antiherpesvirus activity. *Proc. Soc. Exp. Biol. Med.* **199**(2):255–263.

100. Cabral, G. A., Pettit, D. A., and Fischer-Stenger, K. (1993). Marijuana and host resistance to herpesvirus infection. *Adv. Exp. Med. Biol.* **335**:95–106.

101. Lancz, G., Specter, S., and Brown, H. K. (1991). Suppressive effect of delta-9-tetrahydrocannabinol on herpes simplex virus infectivity in vitro. *Proc. Soc. Exp. Biol. Med.* **196**(4):401–404.

102. Lancz, G. *et al.* (1991). Interaction of delta-9-tetrahydrocannabinol with herpesviruses and cultural conditions associated with drug-induced anti-cellular effects. *Adv. Exp. Med. Biol.* **288**:287–304.

103. Medveczky, M. M., Klein, T.W, Friedman, H., and Medveczky, P. G. Delta-9 tetrahydrocannabinol (THC) inhibits lytic replication of gamma oncogenic herpesviruses *in vitro.* BMC Medicine 2004, 2:34 http://www.biomedcentral.com/1741–7015/2/34.

104. Gardella, T. *et al.* (1984). Detection of circular and linear herpesvirus DNA molecules in mammalian cells by gel electrophoresis. *J. Virol.* **50**:248–254.

105. Neyts, J. and De Clercq, E. (1998). In vitro and in vivo inhibition of murine gamma herpesvirus 68 replication by selected antiviral agents. *Antimicrob. Agents Chemother.* **42**:170–172.

106. Usherwood, E. J., Stewart, J. P., and Nash, A. A. (1996). Characterization of tumor cell lines derived from murine gammaherpesvirus-68-infected mice. *J. Virol.* **70**:6516–6518.

107. Goedert, J. *et al.* (2003). Risk factors for Kaposi's sarcoma-associated herpesvirus infection among HIV-1-infected pregnant woman in the USA. *AIDS* **17**:425–433.

    Zwaagstra, J. C. *et al.* (1991). Identification of a major regulatory sequence in the latency associated transcript (LAT) promoter of herpes simplex virus type 1 (HSV-1). *Virology* **182**(1):287–297.

    Batchelor, A. H., Wilcox, K. W. and O'Hare, P. (1994). Binding and repression of the latency-associated promoter of herpes simplex virus by the immediate early 175K protein. *J. Gen. Virol.* **75**(Pt 4):753–767.

    Batchelor, A. H. and O'Hare, P. (1992). Localization of *cis*-acting sequence requirements in the promoter of the latency-associated transcript of herpes simplex virus type 1 required for cell-type-specific activity. *J. Virol.* **66**(6):3573–3582.

4

Cannabinoids and Susceptibility to Neurological Infection by Free-Living Amebae

GUY A. CABRAL and FRANCINE MARCIANO-CABRAL

1. INTRODUCTION

Free-living amebae of the genera *Acanthamoeba* and *Naegleria* have been associated with a variety of human diseases. These amebae have been termed amphizoic since they have the ability to exist as free-living as well as parasitic protozoa.[1] *Acanthamoeba* are causative agents of Granulomatous Amebic Encephalitis (GAE), a fatal chronic protracted progressive disease of the central nervous system (CNS) which also involves the lungs.[2,3] *Naegleria* causes Primary Amebic Meningoencephalitis (PAM), a rapidly fatal disease of the CNS.[4-7] In addition, *Acanthamoeba* is the causative agent of amebic keratitis (AK), a painful sight-threatening disease of the eye.[7-10] However, while PAM occurs in individuals who are immune competent, GAE is generally associated with individuals who suffer from underlying diseases such as malignancies, systemic lupus erythematosus, diabetes, renal failure, cirrhosis, tuberculosis, skin ulcers, human immunodeficiency virus (HIV) infection, or Hodgkin's disease.[7,11-17] Thus, *Acanthamoeba* spp., in contrast to *Naegleria*, act primarily as opportunistic pathogens. Recently, two other free-living amebae from distinct genera, *Balamuthia mandrillaris* and *Sappinia diploidea*, have been associated with CNS

GUY A. CABRAL and FRANCINE MARCIANO-CABRAL • Department of Microbiology and Immunology, Virginia Commonwealth University, Richmond, VA 23298-0678.

Infectious Diseases and Substance Abuse, edited by Herman Friedman *et al.*
Springer, New York, 2005.

infections in humans.[18,19] *B. mandrillaris* was reported to cause fatal amebic encephalitis in both healthy and immune suppressed patients.[20,21]

Acanthamoeba spp. are among the most prevalent protozoa found in the environment.[22] These amebae are distributed worldwide and have been isolated from soil, dust, natural and treated water sources, air-conditioning units, contact lenses and lens cases, eyewash stations, dental treatment units, and hospital and dialysis units.[22] The life cycle of *Acanthamoeba* spp. consists of an actively dividing vegetative trophozoite stage and of a dormant cyst stage (Fig. 1). Trophozoites feed on bacteria, algae, and yeast in the environment by pseudopod formation and phagocytosis or by food-cup formation and ingestion of

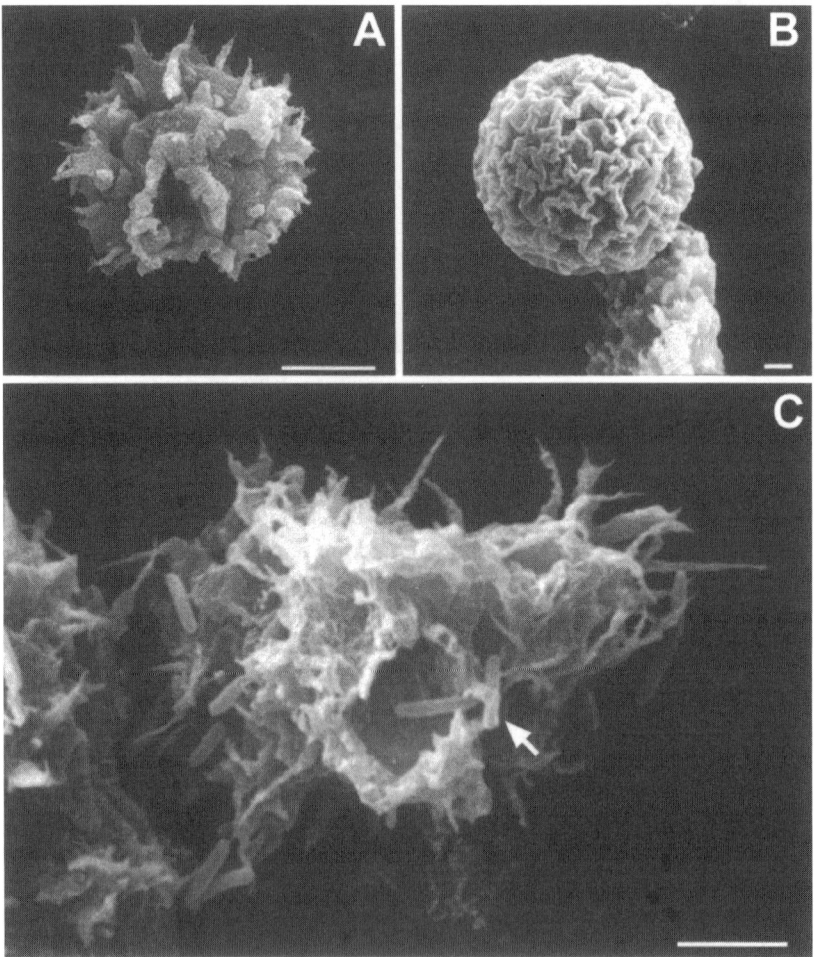

FIGURE 1. Scanning electron micrographs depicting stages in the life cycle of *Acanthamoeba*. (A) Trophozoite of *Acanthamoeba castellanii*. The bar = 10 μm. (B) Cyst of *A. castellanii*. The bar = 1 μm. (C) Trophozoite of *A. castellanii* in the apparent process of ingesting *Escherichia coli* bacteria (arrow). The bar = 20 μm.

particulate matter.[23] Cyst formation occurs under adverse environmental conditions such as food deprivation, dessication, and changes in temperature and pH.[24-26]

2. FREE-LIVING AMEBAE AS OPPORTUNISTIC PATHOGENS

An increasing number of cases of disseminated *Acanthamoeba* infections has been reported in individuals with AIDS.[22] Most of these have been diagnosed postmortem. The clinical course can be fulminant with rapid progression to death. Most patients die in less than 1 month after onset of neurological symptoms.[27-31] In addition, *Acanthamoeba* spp. have been associated increasingly with cutaneous lesions and sinusitis in AIDS patients and in other immunocompromised individuals.[32-38]

The route of infection to the brain is thought to be by inhalation of amebae through the nasal passages and lungs or introduction through skin lesions. Pathological findings are generally of severe hemorrhagic necrosis, fibrin thrombi, and inflammation. The cerebral hemispheres show moderate to severe edema and a chronic inflammatory exudate is observed over the cortex which is comprised mainly of polymorphonuclear leukocytes and mononuclear cells. However, it is unknown whether severe necrosis of the brain is due to direct destruction of tissue by *Acanthamoeba* trophozoites or by induction of inflammatory cytokines such as interleukin 1 (IL-1), tumor necrosis factor (TNF-α), or through the interactive action of both.[39] In addition, dissemination of amebae to other organs such as the liver, kidneys, trachea, and adrenals can occur in immune compromised individuals.[14,33,40,41] Individuals with GAE also may have lung involvement. Trophozoites and cysts have been found in pulmonary alveoli from infected individuals, and pneumonitis is a characteristic feature.[28,40,42-44]

3. THE ROLE OF THE IMMUNE SYSTEM IN *ACANTHAMOEBA* INFECTIONS

Macrophages appear to play an important role in killing *Acanthamoeba*. These cells have been shown to injure amebae and to comprise the major cellular component of granulomas frequently encountered in tissues containing *Acanthamoeba* cysts.[45] Masihi *et al.*[46] studied the effect of the mycobacterial-derived immunopotentiating agents, muramyl dipeptide and trehalose dimycolate, against intranasal *Acanthamoeba* infection in mice. It was found that treatment of mice with these macrophage-activating agents prior to infection protected 40% and 30% of the animals, respectively, to a lethal dose of *Acanthamoeba culbertsoni*. *In vitro* studies using murine macrophages activated *in vivo* with Bacillus Calmétte-Guèrin have demonstrated that activated macrophages are more efficient in injuring *Acanthamoeba* than unstimulated macrophages.[45] Similar results were obtained when unstimulated vs stimulated macrophage-like cells maintained as continuous cell lines were cocultured with *Acanthamoeba.* TNF-α and IL-1α or IL-1β, cytokine products of activated macrophages, were found not to be amebicidal for *Acanthamoeba* either when used alone or in combination.[45] However, hydroxy radicals, hydrogen peroxide, and nitric oxide have been proposed as important amebicidal factors since it has been reported that *Acanthamoeba* are sensitive to hydrogen peroxide.[47] In addition, Stewart *et al.*[48] have reported that rat macrophages, similar to macrophages from mice, chemotax to amebae and kill trophozoites *in vitro.* Thus, although the full range of specific macrophage factors responsible for injuring *Acanthamoeba* has yet to be defined, it is apparent that macrophages activated with immunomodulators are capable of phagocytizing and destroying amebae.[45]

Recent reports indicate that microglial cells, resident macrophages of the brain, also exert amebicidal activity.[39] Microglial cells cocultured with *Acanthamoeba castellanii* were shown to destroy amebae by both phagocytic and lytic processes. Furthermore, *A. castellanii* cocultured with microglial cells induced the production of mRNAs for the cytokines, IL-1α, IL-1β, and TNF-α. Studies of *Acanthamoeba*–microglial interactions have been performed also using highly pathogenic *A. culbertsoni* amebae. Microglial cells cocultured with virulent *A. culbertsoni* exhibited cytopathic changes characteristic of those described for cells undergoing apoptosis while microglial cells cocultured with weakly pathogenic *Acanthamoeba royreba* did not.[49] In view of these observations, it has been postulated that virulent *Acanthamoeba* escape amebicidal activity of macrophages and macrophage-like cells while weakly pathogenic species do not, but rather are lysed or ingested and destroyed by macrophages and macrophage-like cells.[39,45,49]

4. IMMUNOSUPPRESSIVE EFFECTS OF CANNABINOIDS

There is currently a large body of data which indicates that delta-9-tetrahydrocannabinol (THC), the major psychoactive component in marijuana, is immunosuppressive.[50,51] The pharmacological and biological effects of cannabinoids have been attributed to either their lipid solubility and accumulation in membranes or to the stereospecific binding of cannabinoids to a G-protein-coupled receptor which inhibits adenylate cyclase through $G_{i/o}$ proteins.[52,53] To date, two cannabinoid receptor types, CB_1 and CB_2, have been identified.[52,53] CB_1 receptors are present in brain and spinal cord and in certain peripheral tissues while CB_2 receptors are expressed on cells of the immune system. The immunomodulatory effects of THC are thought to be the result of its interaction with receptors, although it has been proposed that effects, also, may be nonreceptor-mediated.[54]

THC has been reported to have a profound effect on the functional state of B- and T lymphocytes, natural killer cells, and macrophages.[50,51] This cannabinoid at concentrations which approach the nanomolar range alters the production of effector molecules by lymphocytes and macrophages, including the expression of Class II molecules of the major histocompatibility complex (MHC) and the elicitation and processing of monokines such as IL-1 and TNF-α.[55] THC has been shown, also, to inhibit mitogen-induced T-cell proliferation and to increase TH2 activity while decreasing TH1 activity of splenocytes.[56] Furthermore, THC has been reported to suppress macrophage-mediated cytotoxicity against tumor cells and extrinsic antiviral activity against virus-infected cells.[57] THC administration to experimental animals, resulted in a dose-related increase in susceptibility to herpes virus infection. Furthermore, this increased susceptibility was shown to correlate with a significant reduction in antibody production and cell-mediated immunity.[58] THC, also, has been reported to augment the susceptibility of mice to infection with the opportunistic pathogen *Legionella pneumophilia*.[59] Increased mortality in these animals was associated with decreased production of interferon gamma (IFN-γ) and increased production of IL-4. It has been suggested, in addition, that marijuana may serve as a cofactor in conjunction with opportunistic pathogens in the progression of infection with retroviruses such as HIV.[60] Recently, alveolar macrophages from marijuana smokers were shown to exhibit impaired ability to ingest and kill *Staphylococcus aureus*.[61] THC has been shown, also, to impair the development of antitumor immunity *in vivo*. Pretreatment of mice with THC for 2 weeks prior to implanting Lewis lung-cancer cells resulted in larger, faster growing tumors which correlated with decreased amounts of the proinflammatory cytokine, IFN-γ, and increased amounts of the immunosuppressive cytokines TGF-β and IL-10.[62]

5. *IN VIVO* EFFECTS OF CANNABINOIDS ON *ACANTHAMOEBA* INFECTION

A $(B_6C_3)F_1$ murine model of amebic encephalitis has been utilized to investigate the effect of THC on immune cells in the brain and the outcome of infection with *Acanthamoeba*. In this model, mice are infected through the intranasal route to mimic one of the natural routes of infection in humans. Trophozoites enter the nasal passages and make their way to the brain from the nasal mucosa following the nerve endings through the cribriform plate. Using this mouse model, a major difference in host resistance to infection was observed between THC-treated and untreated animals infected with *Acanthamoeba*. THC was shown to increase mortalities for mice receiving highly pathogenic ($LD_{50} = 1 \times 10^3$) *A. culbertsoni* (Table I). As anticipated, 50% of vehicle-treated animals expired when inoculated intranasally with a 1 LD_{50} dose of *A. culbertsoni*. In contrast, 85% of mice treated with THC (40 mg/kg) and similarly infected expired. A similar effect of THC on host resistance to *Acanthamoeba* was obtained even for mice inoculated with weakly pathogenic ($LD_{50} = 3 \times 10^6$) *A. castellanii*. THC-treated mice exhibited drug dose-related higher mortalities from infection with *Acanthamoeba* than similarly infected vehicle controls (Table II). A 15% mortality rate was recorded for infected animals treated with vehicle. In contrast, animals receiving 10, 25, or 80 mg/kg THC exhibited approximately 33%, 41%, and 50% mortalities, respectively. Furthermore, *Acanthamoebae* were isolated from brain

TABLE I

**The Effect of THC or Cyclophosphamide (CPA) Treatment on *A. culbertsoni*
Infection in $(B_6C_3)F_1$ Mice[a]**

Treatment[b]	Number of animals	Number dead (%)[c]
None	15	8 (53)
Vehicle	8	4 (50)
CPA 200 mg/kg	8	4 (50)
THC 40 mg/kg	8	7 (85)

[a]Reproduced with permission of *J. Eukaryotic Microbiol.*[(63)]
[b]Mice (3 weeks old) were injected intraperitoneally with vehicle ethanol : emulphor : saline, 1 : 1 : 18, CPA or
THC prior to intranasal inoculation with 1×10^3 *A. culbertsoni*.
[c]Mice were observed for 30 days postinoculation and the number of dead animals was recorded.

TABLE II

**The Effect of THC on Amebic Encephalitis Caused by
A. castellanii in $(B_6C_3)F_1$ Mice**

Treatment[a]	Number of mice	Number dead (%)[b]	Amebae[c]	% survivors with amebae[d]
Vehicle	20	3 (15)	Brain	0
10 mg/kg THC	12	4 (33)	Brain, lungs	0
25 mg/kg THC	12	5 (41)	Brain, lungs	0
80 mg/kg THC	12	6 (50)	Brain, lungs	100
200 mg/kg CPA[e]	12	0 (0)	None	0

[a]Female mice (3 weeks old) were injected ip with vehicle (ethanol : emulphor : saline, 1 : 1 : 18),
cyclophosphamide (CPA) or THC prior to inoculation via the intranasal route with 1×10^6 *A. castellanii*.
[b]The number in brackets represents the percentage of animals which died over a 30-day period.
[c]At the time of death, brain and lungs were removed and cultured. Amebae were cultured from the respective
organs.
[d]Mice which survived the 30-day observation period were sacrificed and examined for the presence of amebae
in brain and lungs.
[e]200 mg/kg of CPA was injected ip one day prior to challenge with *Acanthamoeba*.

tissue as well as from lungs of all animals that died indicating colonization at
multiple sites.[(63)] Cyclophosphamide (CPA), a potent immunosuppressive agent
which targets B lymphocytes, had no discernible effect on host resistance
to either *A. castellanii* or *A. culbertsoni* infection. In fact, for animals inoculated
with the weakly pathogenic *A. castellanii*, no mortalities were recorded for mice
receiving 200 mg/kg CPA.

The greater severity of disease for mice treated with THC occurred concurrently with dysfunction in responsiveness of macrophage-like cells to
Acanthamoeba in the brain. Staining of paired serial sections of brain from infected
mice treated with THC with anti-Mac-1 or anti-*Acanthamoeba* antibodies demonstrated that Mac-1 + cells were abundant in focal areas of infected tissue for vehicle-treated animals (Fig. 2). Few amebae were colocalized in focal areas replete
with Mac-1 + cells. In contrast, foci in tissue from infected, THC-treated mice contained many amebae but few Mac-1 + cells. This paucity of Mac-1 + cells at focal
sites of *Acanthamoeba* infection suggests that macrophage-like cells either do not

migrate to infected areas or, alternatively, are destroyed by the *Acanthamoeba*. These results are consistent with the *in vivo* data indicating that CPA, which targets B lymphocytes, had a minimal effect on *Acanthamoeba* infection of mice. The colocalization of Mac1+ cells and *Acanthamoeba* in brains of vehicle-treated mice

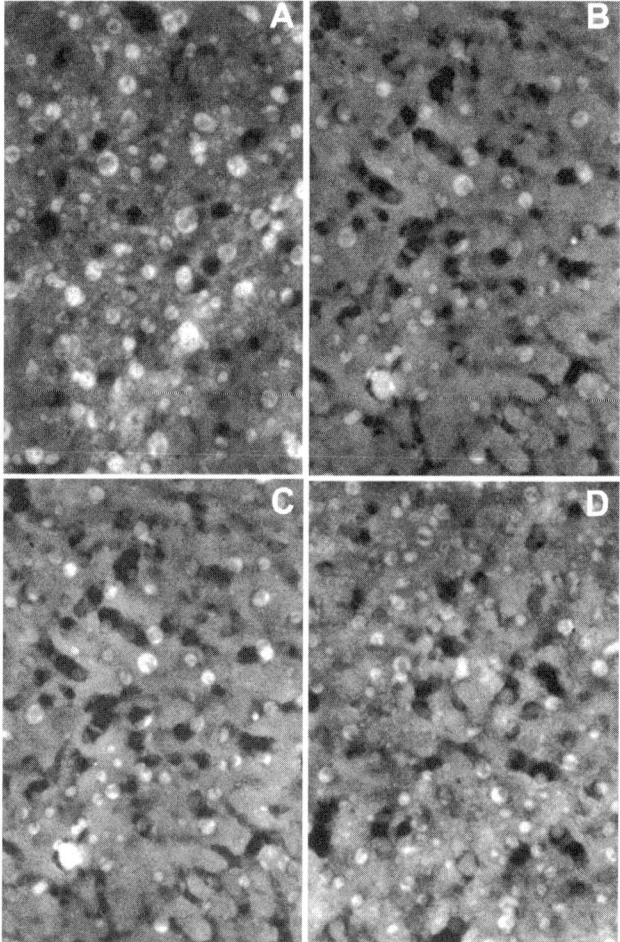

FIGURE 2. Immunofluorescence microscopy of cryostat sections of brain from vehicle-treated and THC-treated *Acanthamoeba*-infected $(B_6C_3)F_1$ mice at 8 days postintranasal exposure. (A) Indirect immunofluorescence staining of brain sections from vehicle-treated (ethanol : emulphor : saline, 1 : 1 : 18) mice using anti-Mac1 as the primary antibody reveals aggregation of numerous Mac1+ cells at focal sites containing *Acanthamoeba*. (B) Indirect immunofluorescence staining of a paired serial section of brain from the same vehicle-treated animal using anti-*Acanthamoeba* antibody reveals the presence of a relatively low number of amebae colocalized with the Mac1+ cells. (C) Indirect immunofluorescence staining of brain sections from THC-treated (40 mg/kg) mice using anti-Mac1 as the primary antibody reveals aggregation of a relatively low number of Mac1+ cells at focal sites containing *Acanthamoeba*. (D) Indirect immunofluorescence staining of a paired serial section of brain from the same THC-treated animal using anti-*Acanthamoeba* antibody reveals the presence of a relatively high number of amebae colocalized with the Mac1+ cells. All micrographs ×70.

was observed to occur at a relatively early phase of infection (i.e., 8 days postin-tranasal exposure). These results suggest that the Mac1+ cells represent microglia, the resident macrophages of the brain, rather than macrophages intro-duced from nonneuronal peripheral sites. This postulate is supported by the observation that cells exhibiting the T-cell phenotypic marker Thy1.2 were not detected in paired sections from vehicle-treated or THC-treated *Acanthamoeba*-exposed mice (Fig. 3). The absence of these cells is consistent with an intact brain–blood barrier at this stage of infection (i.e., 8 days postinoculation), at least from the perspective of a state of dissolution wherein immune cells from the periphery such as monocytes and lymphocytes would be invasive of brain tissue.

Indeed, recent reports indicate that microglia exert amebicidal activity. Marciano-Cabral *et al.*[39] reported that microglial cells co-cultured with *A. castellanii* destroyed amebae by both phagocytic and lytic processes. Furthermore, *A. castellanii* co-cultured with microglia induced the production of mRNAs for the cytokines IL-1α, IL-1β, and TNF-α by these cells of macrophage lineage. In addition, microglia co-cultured with the highly pathogenic *A. culbertsoni* exhib-ited cytopathic changes characteristic with those described for cells undergoing apoptosis while microglial cells co-cultured with weakly pathogenic *A. royreba* did not.[49] In view of these observations, it has been postulated that virulent

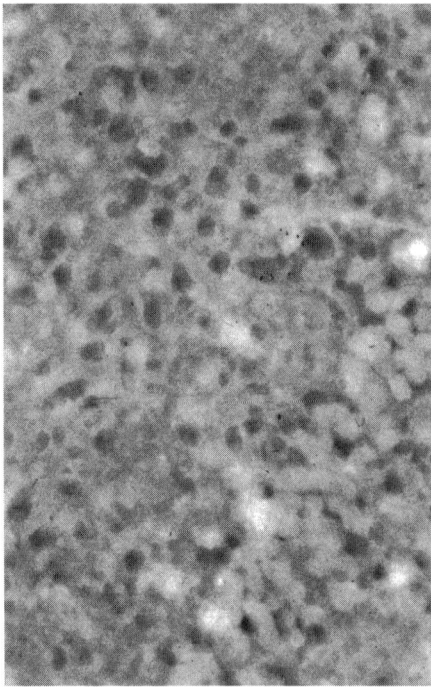

FIGURE 3. Immunofluorescence microscopy of cryostat section of brain from vehicle-treated *Acanthamoeba*-infected $(B_6C_3)F_1$ mouse at 8 days postintranasal exposure incubated with anti-Thy1.2 antibody. Focal areas of colocalized Mac1+ cells and *Acanthamoeba* exhibit a paucity of cells containing the phenotypic marker for T lymphocytes.

Acanthamoeba escape amebicidal activity of macrophages and macrophage-like cells while weakly pathogenic species do not but rather are lysed or ingested and destroyed.[39,45,49] In this context, it is possible that THC may exacerbate the process wherein *Acanthamoeba* escape amebicidal activity.

6. *IN VITRO* EFFECTS OF CANNABINOIDS ON MICROGLIAL RESPONSE TO *ACANTHAMOEBA*

In vivo studies have implicated microglia as exhibiting altered responsiveness to brain infection with *Acanthamoeba*. In order to extend these studies on effects of THC on macrophage-like cell activity, *in vitro* coculture experiments have been performed. *Acanthamoeba* were shown to elicit gene expression for the proinflammatory cytokines, IL-1α, IL-1β, IL-6, and TNF-α. However, the most robust induction was that for IL-1α and IL-1β. THC treatment (10^{-6}M and 10^{-5}M) of microglia antecedent to exposure to *Acanthamoeba* resulted in decreases in levels of mRNAs for these two cytokines (Figs 4 and 5). Maximal effect in terms of decreased levels of IL-1α and IL-1β mRNAs was observed for microglia pretreated with 10^{-5}M THC.

These observations are consistent with those obtained from previous studies which have indicated that cannabinoids alter the expression of cytokines elicited by microglia. However, it is becoming apparent that the extent and intensity of the effect of cannabinoids on cytokine production may depend on the type of inducing agent which initiated the cytokine response. For example, bacterial lipopolysaccharide (LPS) has been shown to elicit a robust induction of mRNAs not only for the proinflammatory cytokines IL-1α and IL-1β but also for IL-6 and TNF-α. In contrast to results obtained following exposure to *Acanthamoeba*, THC, as well as the highly potent synthetic cannabinoid agonist CP55940, exerted a dose-related inhibition in LPS-inducible mRNA production of IL-6 and TNF-α.[64] THC and CP55940 also have been reported to inhibit the production of inducible nitric oxide by microglia in response to LPS used in concert with IFN-γ.[65] In this context, it is reasonable to anticipate that highly pathogenic amebae such as *Naegleria fowleri*, which cause a rapidly fatal acute disease, and *Acanthamoeba*, which induce a chronic protracted progressive disease, may elicit proinflammatory (or even anti-inflammatory) cytokine responses for which patterns, as well as robustness of expression of different cytokines, may be distinctive. Under such circumstances, cannabinoids may exhibit differential effects on cytokine responsiveness to species of free-living amebae which elicit distinctive disease patterns. Assessment of the effects of cannabinoids on immune elements resident in the brain which respond to infectious agents is in its infancy. Studies related to this issue are the focus of investigation in this and other laboratories.

7. SUMMARY AND CONCLUSIONS

There is accumulating evidence that free-living amebae such as *Acanthamoeba* pose a health risk to individuals, particularly those who suffer from

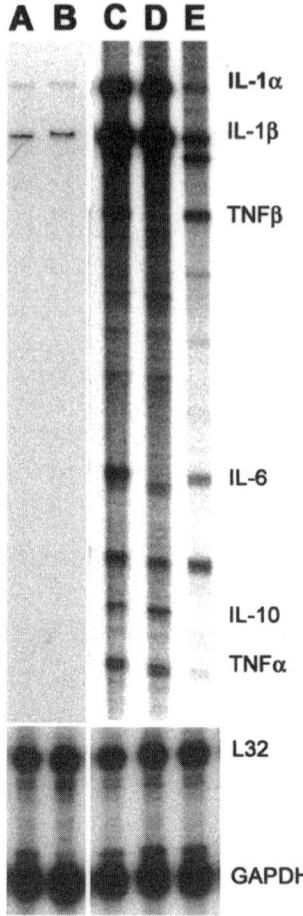

FIGURE 4. Ribonuclease protection assay demonstrating differential effect of THC on production of proinflammatory cytokine mRNAs by microglia cocultured with *A. castellanii* (American Type Culture Collection, 50494). Purified neonatal rat brain cortical microglia (1×10^6 cells) were maintained in medium alone, medium containing vehicle (0.1% ethanol), or medium containing THC (10^{-6}M) or 10^{-5}M for 3 hr. Microglial cultures then were either inoculated with *A. castellanii* (1×10^5) or placebo (medium), and were maintained for an additional 6 hr. Total RNA was isolated from cultures, and cytokine mRNA species were detected using the RiboQuant™ rCK-1 template set RNase Protection Assay (Pharmingen, San Diego, CA) according to the manufacturer's instructions. The probes were made with [^{32}P]UTP with a specific activity of greater than 3,000 Ci/mmol. The RNA samples were hybridized to the rCK-1 probe set overnight at 56°C. The protected fragments were subjected to RNase digestion, resolved on a 6% polyacrylamide gel containing 6M urea and imaged using XOMAT-AR film (Rochester, NY). The pixel intensity of each band was quantified using a Molecular Dynamics 445SI Phosphoimager with the Image Quant 4.1 software (Molecular Dynamics, Sunnyvale, CA). The amount of cytokine mRNA was normalized for loading by dividing the pixel value for the cytokine band by the sum of the pixel values for the housekeeping gene mRNA for GAPDH. Relatively low levels of cytokine mRNAs were detected from microglial cultures maintained in (A) medium alone or (B) medium containing vehicle. (C) Relatively high levels of mRNAs for cytokines were elicited in vehicle-treated cultures in response to *Acanthamoeba*. Note that particularly high levels of IL-1α and IL-1β mRNAs were produced. Decreases in levels of cytokine mRNAs, particularly for IL-1α and IL-1β, were noted for microglial cultures treated with (D) 10^{-6}M THC or (E) 10^{-5}M THC.

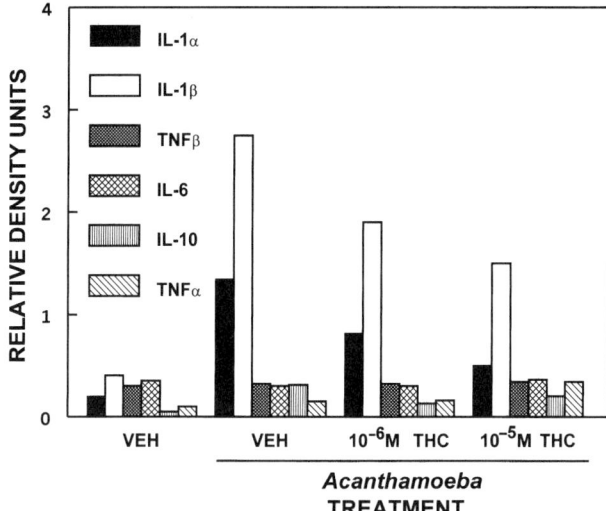

FIGURE 5. Graphic representation of ribonuclease protection assay from a second experiment demonstrating differential effect of THC on production of proinflammatory cytokine mRNAs by microglia cocultured with *A. castellanii*. Minimal levels of mRNAs for IL-1α, IL-1β, and TNF-α were detected for microglia maintained in medium alone. Note the relatively high levels of mRNA for IL-1β from total RNA of microglia cultured with *Acanthamoeba*. THC treatment (10^{-6}M–10^{-5}M) resulted in a decrease in these levels.

a variety of immune deficiencies such as AIDS. Because marijuana is the most widely used illicit substance, it is rational to anticipate that a subset of immune compromised individuals also uses marijuana either in a recreational or in a self-administered therapeutic mode. Reports that cannabinoids such as THC alter host responsiveness to a variety of infectious agents including free-living amebae such as *Acanthamoeba* indicate that individuals who suffer from immune deficiencies such as AIDS patients may be especially susceptible to *Acanthamoeba* infections. Furthermore, the recognition that many neuropathies are characterized by a state of persistent production of proinflammatory cytokines, and that cannabinoids may alter their expression, indicates a potential for these compounds to alter the outcome of disease. However, there is only a limited number of studies which have examined the effects of cannabinoids on brain infections. It is possible that for chronic, persistent amebic infection, cannabinoids may exert effects distinctive from those associated with rapid acute amebic infection. In addition, the role of cannabinoid receptors in cannabinoid-mediated alteration of macrophage-like function related to neurological infection by amebae remains to be defined. While it has been demonstrated that microglia express both CB_1 and CB_2 receptors,[65,66] and that cannabinoid-mediated inhibition of inducible nitric oxide is linked functionally to the CB_1 receptor,[65] a role of a CB_2 receptor in cannabinoid-mediated events has not been established. However, it has been shown *in vitro* that levels of the CB_2 receptor are modulated in relation to cell activation state[66] and that highest levels are expressed when microglia are in their ameboid "responsive" state. Thus, functional activities associated with this activation state may be the most sensitive to the action of cannabinoids. Indeed, "responsive" microglia exhibit characteristic functional features which include chemotaxis and migration toward sites of infection and phagocytosis of "foreign" particulates. Studies indicating that THC administered to mice resulted in inhibition of localization of Mac1+ cells at focal sites in the brain containing *Acanthamoeba* are consistent with this postulate. Finally, recent studies have indicated that a third so-called "non-CB_1, non-CB_2" cannabinoid receptor may be present in the brain.[67] Thus, a picture is emerging concerning the existence of a diverse network of cannabinoid-linked signal transductional pathways in the brain which may play a role in the modulation of a disparate array of immune functional activities.

ACKNOWLEDGMENTS. This work was supported in part by awards from the National Institute on Drug Abuse, National Institutes of Health: DA05274, DA05832, DA09789, and DA15608.

REFERENCES

1. Page, F. C. (1974). *Rosculus ithacus* Hawes 1963 (Amoebida, Flabelluidae) and the amphizoic tendency in amoebae. *Acta Protozool.* **13:**143.
2. Martinez, A. J., Sotelo-Avila, C., Garcia-Tamayo, J., Moron, J. T., Willaert, E., and Stamm, W. P. (1977). Meningoencephalitis due to *Acanthamoeba* SP. Pathogenesis and clinico-pathological study. *Acta Neuropathol. (Berl)* **37:**183.
3. Duma, R. J., Helwig, W. B., and Martinez, A. J. (1978). Meningoencephalitis and brain abscess due to a free-living amoeba. *Ann. Intern. Med.* **88:**468.
4. Butt, C. G. (1966). Primary amebic meningoencephalitis. *N. Engl. J. Med.* **274:**1473.
5. Callicott, J. H., Jr. (1968). Amebic meningoencephalitis due to free-living amebas of the *Hartmannella (Acanthamoeba)-Naegleria* group. *Am. J. Clin. Pathol.* **49:**84.
6. Marciano-Cabral, F. (1988). Biology of *Naegleria* spp. *Microbiol. Rev.* **52:**114.
7. Martinez, A. J. and Visvesvara, G. S. (1997). Free-living, amphizoic and opportunistic amebas. *Brain Pathol.* **7:**583.
8. Culbertson, C. G. (1961). Pathogenic *Acanthamoeba (Hartmannella)*. *Am. J. Clin. Pathol.* **35:**195.
9. Nagington, J., Watson, P. G., Playfair, T. J., McGill, J., Jones, B. R., and Steele, A. D. (1974). Amoebic infection of the eye. *Lancet* **2:**1537.
10. Jones, B. R., McGill, J. I., and Steele, A. D. (1975) Recurrent suppurative kerato-uveitis with loss of eye due to infection by *Acanthamoeba castellani*. *Trans. Ophthalmol. Soc. UK.* **95:**210.
11. Jager, B. V. and Stamm, W. P. (1972). Brain abscesses caused by free-living amoeba probably of the genus *Hartmannella* in a patient with Hodgkin's disease. *Lancet* **2:**1343.
12. Willaert, E., Stevens, A. R., and Healy, G. R. (1978). Retrospective identification of *Acanthamoeba culbertsoni* in a case of amoebic meningoencephalitis. *J. Clin. Pathol.* **31:**717.
13. Grunnet, M. L., Cannon, G. H., and Kushner, J. P. (1981). Fulminant amebic meningo-encephalitis due to *Acanthamoeba*. *Neurology* **31:**174.
14. Martinez, A. J. (1982). Acanthamoebiasis and immunosuppression. Case report. *J. Neuropathol. Exp. Neurol.* **41:**548.

15. Harwood, C. R., Rich, G. E., McAleer, R., and Cherian, G. (1988). Isolation of *Acanthamoeba* from a cerebral abscess. *Med. J. Aust.* 148:47.

16. Koide, J., Okusawa, E., Ito, T., Mori, S., Takeuchi, T., Itoyama, S. *et al.* (1998). Granulomatous amoebic encephalitis caused by *Acanthamoeba* in a patient with systemic lupus erythematosus, *Clin. Rheumatol.* 17:329.

17. Feingold, J. M., Abraham, J., Bilgrami, S., Ngo, N. Visvesara, G. S., Edwards, R. L. *et al.* (1998). *Acanthamoeba* meningoencephalitis following autologous peripheral stem cell transplantation, *Bone Marrow Transplant.* 22:297.

18. Visvesvara, G. S., Schuster, F. L., and Martinez, A. J. (1993). *Balamuthia mandrillaris*, N. G., N. Sp., agent of amebic meningoencephalitis in humans and other animals. *J. Eukaryot. Microbiol.* 40:504.

19. Gelman, B. B., Rauf, S. J., Nader, R., Popov, V., Borkowski, J., Chaljub, G. *et al.* (2001). Amoebic encephalitis due to *Sappinia diploidea*, YAMA 285:2450.

20. Martinez, A. J., Guerra, A. E., Garcia-Tamayo, J., Cespedes, G., Gonzalez-Alfonzo, J. E., and Visvesvara, G. S. (1994). Granulomatous amebic encephalitis: A review and report of a spontaneous case from Venezuela. *Acta Neuropathol.* (*Berl*) 87:430.

21. Rowen, J. L., Doerr, C. A., Vogel, H., and Baker, C. J. (1995). *Balamuthia mandrillaris*: A newly recognized agent for amebic meningoencephalitis. *Pediatr. Infect. Dis. J.* 14:705.

22. Marciano-Cabral, F. and Cabral, G. (2003). The importance of *Acanthamoeba* spp. as agents of human disease, *Clin. Microbiol. Rev.* 16:273

23. Pettit, D. A., Williamson, J., Cabral, G. A., and Marciano-Cabral, F. (1996). In vitro destruction of nerve cell cultures by *Acanthamoeba* spp.: A transmission and scanning electron microscopy study. *J. Parasitol.* 82:769.

24. Bowers, B. and Korn, E. D. (1969). The fine structure of *Acanthamoeba castellanii* (Neff strain). II. Encystment. *J. Cell Biol.* 41:786.

25. Chagla, A. H. and Griffiths, A. J. (1974). Growth and encystation of *Acanthamoeba castellanii*. *J. Gen. Microbiol.* 85:139.

26. Byers, T. J., Akins, R. A., Maynard, B. J., Lefken, R. A., and Martin, S. M. (1980). Rapid growth of *Acanthamoeba* in defined media; induction of encystment by glucose–acetate starvation. *J. Protozool.* 27:216.

27. Wiley, C. A., Safrin, R. E., Davis, C. E., Lampert, P. W., Braude, A. I., Martinez, A. J. *et al.* (1987). *Acanthamoeba* meningoencephalitis in a patient with AIDS. *J. Infect. Dis.* 155:130.

28. Gardner, H. A., Martinez, A. J., Visvesvara, G. S., and Sotrel, A. (1991). Granulomatous amebic encephalitis in an AIDS patient. *Neurology* 41:1993.

29. Gordon, S. M., Steinberg, J. P., DuPuis, M. H., Kozarsky, P. E., Nickerson, J. F., and Visvesvara, G. S. (1992). Culture isolation of *Acanthamoeba* species and leptomyxid amebas from patients with amebic meningoencephalitis, including two patients with AIDS. *Clin. Infect. Dis.* 15:1024.

30. Tan, B., Weldon-Linne, C. M., Rhone, D. P., Penning, C. L., and Visvesvara, G. S. (1993). *Acanthamoeba* infection presenting as skin lesions in patients with the acquired immunodeficiency syndrome. *Arch. Pathol. Lab Med.* 117:1043.

31. Calore, E. E., Cavaliere, M. J., and Calore, N. M. (1997). Cerebral amebiasis in the acquired immunodeficiency syndrome. *Acta Neurol. Belg.* 97:248.

32. Gullett, J., Mills, J., Hadley, K., Podemski, B., Pitts, L., and Gelber, R. (1979). Disseminated granulomatous acanthamoeba infection presenting as an unusual skin lesion. *Am. J. Med.* 67:891.

33. Martinez, A. J. and Janitschke, K. (1985). *Acanthamoeba*, an opportunistic microorganism: A review. *Infection* 13:251.

34. Friedland, L. R., Raphael, S. A., Deutsch, E. S., Johal, J., Martyn, L. J., Visvesvara, G. S. *et al.* (1992). Disseminated *Acanthamoeba* infection in a child with symptomatic human immunodeficiency virus infection. *Pediatr. Infect. Dis. J.* 11:404.

35. May, L. P., Sidhu, G. S., and Buchness, M. R. (1992). Diagnosis of *Acanthamoeba* infection by cutaneous manifestations in a man seropositive to HIV. *J. Am. Acad. Dermatol.* 26:352.

36. Helton, J., Loveless, M., and White, C. R., Jr. (1993). Cutaneous acanthamoeba infection associated with leukocytoclastic vasculitis in an AIDS patient. *Am. J. Dermatopathol.* 15:146.

37. Dunand, V. A., Hammer, S. M., Rossi, R., Poulin, M., Albrecht, M. A., Doweiko, J. P. *et al.* (1997). Parasitic sinusitis and otitis in patients infected with human immunodeficiency virus: Report of five cases and review. *Clin. Infect. Dis.* **25**:267.

38. Torno, M. S., Jr., Babapour, R., Gurevitch, A., and Witt, M. D. (2000). Cutaneous acanthamoebiasis in AIDS. *J. Am. Acad. Dermatol.* **42**:351.

39. Marciano-Cabral, F., Puffenbarger, R., and Cabral, G. A. (2000). The increasing importance of *Acanthamoeba* infections. *J. Eukaryot. Microbiol.* **47**:29.

40. Khalife, G. E., Pambuccian, S. W., Visvesvara, G. S., and Horten, B. (1994). Disseminated acanthamoeba infection masquerading as bacillary angiomatosis in a patient with AIDS. *Int. J. Surg. Pathol.* **2**:11.

41. Murakawa, G. J., McCalmont, T., Altman, J., Telang, G. H., Hoffman, M. D., Kantor, G. R. *et al.* (1995). Disseminated acanthamebiasis in patients with AIDS. A report of five cases and a review of the literature. *Arch. Dermatol.* **131**:1291.

42. Visvesvara, G. S., Mirra, S. S., Brandt, F. H., Moss, D. M., Mathews, H. M., and Martinez, A. J. (1983). Isolation of two strains of *Acanthamoeba castellanii* from human tissue and their pathogenicity and isoenzyme profiles. *J. Clin. Microbiol.* **18**:1405.

43. Martinez, A. J. (1987). Clinical manifestations of free-living amebic infections. In E. G. Rondanelli (ed.), *Amphizoic Amoebae: Human Pathology*, Piccin Nuova Libraria, Padua.

44. Heffler, K. F., Eckhardt, T. J., Reboli, A. C., and Stieritz, D. (1996). *Acanthamoeba endophthalmitis* in acquired immunodeficiency syndrome. *Am. J. Ophthalmol.* **122**:584.

45. Marciano-Cabral, F. and Toney, D. M. (1998). The interaction of *Acanthamoeba* spp. with activated macrophages and with macrophage cell lines. *J. Eukaryot. Microbiol.* **45**:452.

46. Masihi, K. N., Bhaduri, C. R., Werner, H., Janitschke, K., and Lange, W. (1986). Effects of muramyl dipeptide and trehalose dimycolate on resistance of mice to *Toxoplasma gondii* and *Acanthamoeba culbertsoni* infections. *Int. Arch. Allergy Appl. Immunol.* **81**:112.

47. Ferrante, A. (1991). Immunity to *Acanthamoeba*. *Rev. Infect. Dis.* **13**(Suppl 5):S403.

48. Stewart, G. L., Kim, I., Shupe, K., Alizadeh, H., Silvany, R., McCulley, J. P. *et al.* (1992). Chemotactic response of macrophages to *Acanthamoeba castellanii* antigen and antibody-dependent macrophage-mediated killing of the parasite. *J. Parasitol.* **78**:849.

49. Shin, H. J., Cho, M. S., Kim, H. I., Lee, M., Park, S., Sohn, S. *et al.* (2000). Apoptosis of primary-culture rat microglial cells induced by pathogenic *Acanthamoeba* spp. *Clin. Diagn. Lab Immunol.* **7**:510.

50. Klein, T. W., Friedman, H., and Specter, S. (1998). Marijuana, immunity and infection. *J. Neuroimmunol.* **83**:102.

51. Cabral, G. A. and Dove Pettit, D. A. (1998). Drugs and immunity: Cannabinoids and their role in decreased resistance to infectious disease. *J. Neuroimmunol.* **83**:116.

52. Matsuda, L. A., Lolait, S. J., Brownstein, M. J., Young, A. C., and Bonner, T. I. (1990). Structure of a cannabinoid receptor and functional expression of the cloned cDNA. *Nature* **346**:561.

53. Munro, S., Thomas, K. L., and Abu-Shaar, M. (1993). Molecular characterization of a peripheral receptor for cannabinoids. *Nature* **365**:61.

54. Felder, C. C., Veluz, J. S., Williams, H. L., Briley, E. M., and Matsuda, L. A. (1992). Cannabinoid agonists stimulate both receptor- and non-receptor-mediated signal transduction pathways in cells transfected with and expressing cannabinoid receptor clones. *Mol. Pharmacol.* **42**:838.

55. Fischer-Stenger, K., Dove Pettit, D. A., and Cabral, G. A. (1993). Delta 9-tetrahydrocannabinol inhibition of tumor necrosis factor-alpha: Suppression of post-translational events. *J. Pharmacol. Exp. Ther.* **267**:1558.

56. Klein, T. W., Newton, C., Zhu, W., Nakachi, N., and Friedman, H. (2000). Delta 9-tetrahydrocannabinol treatment suppresses immunity and early IFNγ, IL-12, and IL-12 receptor B2 responses to *Legionella pneumophila* infection. *J. Immunol.* **164**:6461.

57. Cabral, G. A. and Vasquez, R. (1992). Delta 9-Tetrahydrocannabinol suppresses macrophage extrinsic antiherpesvirus activity. *Proc. Soc. Exp. Biol. Med.* **199**:255.

58. Mishkin, E. M. and Cabral, G. A. (1985). Delta-9-Tetrahydrocannabinol decreases host resistance to herpes simplex virus type 2 vaginal infection in the B6C3F1 mouse. *J. Gen. Virol.* **66**(Pt 12):2539.

59. Newton, C., Klein, T. W., and Friedman, H. (1994). Secondary immunity to *Legionella pneumophila* and Th1 activity are suppressed by delta-9-tetrahydrocannabinol injection. *Infec. Immun.* **62:**4015.

60. Specter, S., Lancz, G., Westrich, G., and Friedman, H. (1991). Delta-9-tetrahydrocannabinol augments murine retroviral induced immunosuppression and infection. *Int. J. Immunopharmacol.* **13:**411.

61. Baldwin, G. C., Tashkin, D. P., Buckley, D. M., Park, A. N., Dubinett, S. M., and Roth, M. D. (1997). Marijuana and cocaine impair alveolar macrophage function and cytokine production. *Am. J. Respir. Crit Care Med.* **156:**1606.

62. Zhu, L. X., Sharma, S., Stolina, M., Gardner, B., Roth, M. D., Tashkin, D. P., and Dubinett, S. M. (2000). Delta-9-tetrahydrocannabinol inhibits antitumor immunity by a CB2 receptor-mediated, cytokine-dependent pathway. *J. Immunol.* **165:**373.

63. Marciano-Cabral, F., Ferguson, T., Bradley, S. G., and Cabral, G. (2001). Delta-9-tetrahydrocannabinol (THC), the major psychoactive component of marijuana, exacerbates brain infection by Acanthamoeba. *J. Eukaryot. Microbiol.* Suppl:4S.

64. Puffenbarger, R. A., Boothe, A. C., and Cabral, G. A. (2000). Cannabinoids inhibit LPS-inducible cytokine mRNA expression in rat microglial cells. *Glia* **29:**58.

65. Waksman, Y., Olson, J. M., Carlisle, S. J., and Cabral, G. A. (1999). The central cannabinoid receptor (CB1) mediates inhibition of nitric oxide production by rat microglial cells. *J. Pharmacol. Exp. Ther.* **288:**1357.

66. Carlisle, S. J., Marciano-Cabral, F., Staab, A., Ludwick, C., and Cabral, G. A. (2002). Differential expression of the CB2 cannabinoid receptor by rodent macrophages and macrophage-like cells in relation to cell activation. *Int. Immunopharmacol.* **2:**69.

67. Breivogel, C. S., Griffin, G., Di, M. V., and Martin, B. R. (2001). Evidence for a new G protein-coupled cannabinoid receptor in mouse brain. *Mol. Pharmacol.* **60:**155.

5

Legionella Infection and Cannabinoids

HERMAN FRIEDMAN, CATHERINE NEWTON, and
THOMAS W. KLEIN

1. INTRODUCTION

The opportunistic intracellular bacterial pathogen *Legionella pneumophila* is ubiquitous and widely distributed in the environment, including air-conditioners' cooling towers and recirculating warm or hot water plumbing in homes, industries, and institutions. This organism was first isolated and identified in 1976 following the outbreak of pneumonia among American legionnaires attending an annual convention in a downtown Philadelphia hotel, where approximately 200 of the 3,000 conventioneers developed a serious pneumonia, with about 25–30 deaths. Epidemiologists from CDC found that attendees at this convention were exposed to this bacterium from vents contaminated from the air-conditioning cooling towers located on the roof of the hotel. Epidemiologic studies revealed that those who developed pneumonia, especially those who succumbed to the disease, were less competent immunologically than the average conventioneer. Numerous subsequent studies established that a deficiency of normal immune responsiveness was an important risk factor for increased susceptibility, including individuals given immunosuppressive drugs for organ transplantation or those with an immunodeficiency disease such as AIDS.

In the past two decades it has been well established that immunity to Legionella, which preferentially infects monocytes/macrophages, depends

HERMAN FRIEDMAN, CATHERINE NEWTON, and THOMAS W. KLEIN • Department of Medical Microbiology and Immunology, University of South Florida College of Medicine, Tampa, FL 33612.

Infectious Diseases and Substance Abuse, edited by Herman Friedman *et al.*
Springer, New York, 2005.

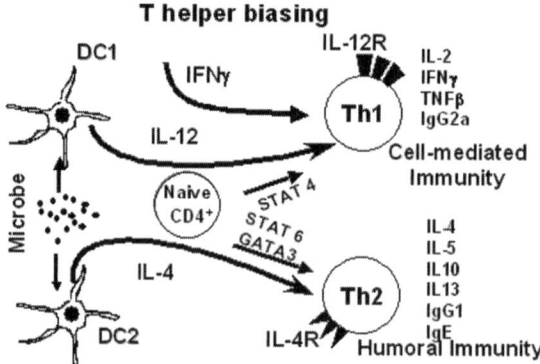

FIGURE 1. Biasing of T-helper cell development. Legionella stimulate different dendritic cell (DC) types to produce cytokines that cause naïve precursors of CD4+ T-helper cells to differentiate into either T-helper 1 (Th1) or T-helper 2 (Th2) cells. These cells then regulate either cell-mediated or humoral immunity, cytokines, interferon gamma (IFN-γ), interleukin (IL), IL-12 receptor (IL-4R), signal transducers and activators of transcription 4 and 6 (STAT 4 and 6), GATA nucleotide sequence (GATA 3), tumor necrosis factor beta (TNF-β), or immunoglobulin (Ig) (with permission from ref. [5]).

primarily on cell-mediated immunity associated with activated T cells and immunoregulatory cytokines, including interleukins and interferons as well as proinflammatory cytokines such as TNF-α.[1–3] Studies in our laboratory as well as others have shown that marijuana cannabinoids, such as delta-9-tetrahydrocannabinol (THC), have marked effects on the immune response system and can suppress T-cell based immunity and cytokines produced by immunoregulatory immune cells.[1,4–6] Reports from our laboratory showed that rodents treated with THC develop heightened susceptibility to *L. pneumophila* infection.[7,6] A number of studies previously had shown that THC enhanced susceptibility to bacterial endotoxins and infection by viruses.[8–14] Furthermore, studies concerning possible mechanisms involved showed that THC injection of mice suppressed Th1 immunity (see Fig. 1) important in resistance to Legionella infection, inhibiting mobilization of IFN-γ as well as IL-12 and increasing production of the Th2 promoting cytokine IL-4.[6,14–18]

2. THC EFFECTS ON IMMUNE CELLS AND CYTOKINE PRODUCTION

Macrophages are an important cell type for host defense mechanisms, especially against infectious agents, since these cells engage and eliminate foreign substances, including microorganisms. Also, as antigen-presenting and cytokine-secreting cells, they are uniquely positioned to regulate the immune response. In the lungs they represent a constant barrier to pulmonary infections, including Legionella infection. Various studies indicated that smoking marijuana significantly affected pulmonary alveolar macrophages, thus compromising host defenses.[19,20] Pulmonary alveolar macrophages from rodents obtained from pulmonary lavage were depressed by marijuana smoke with regard to bactericidal capacity against *Staphylococcus albus*. Furthermore, THC affected the function of mouse macrophages *in vitro*, including peritoneal macrophages.[5,17,21,22] The marijuana components THC or cannabidiol induced in peritoneal macrophages *in vitro* a pattern of vacuolation similar to that seen in cells from the lungs of hashish smokers.[20,23,24] Studies also showed that relatively low amounts of THC significantly altered the ability of macrophages to spread in culture, an indication of normal function.[22] Similarly, rodent macrophage cultures treated with THC had altered antigen-processing ability, as well as the ability to produce cytokines which activate lymphocytes.[25]

A number of studies also showed that marijuana components, especially THC, directly inhibited lymphocyte proliferative responses to both T- and B-cell mitogens as well as to bacterial products such as endotoxins (see Fig. 2). Furthermore, THC also suppressed the proliferative response to mitogens such as Con A and PHA, indicating that both T- and B-lymphocytes are susceptible to the suppressive effects.[18,22,26–28] The proliferative response of murine B cells to bacterial LPS appeared to be most suppressed. Besides affecting proliferation of lymphocytes, THC inhibits the ability of lymphocyte cultures to produce the important cytokines IFN-γ and interleukins necessary for a productive antimicrobial response. Many of these effects of THC and other marijuana components

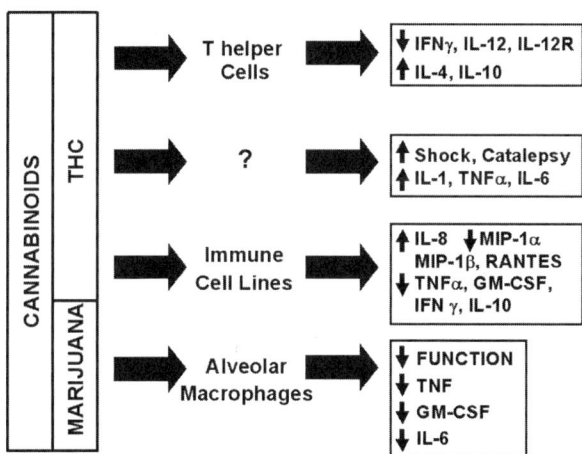

FIGURE 2. Marijuana and cannabinoids modulate cytokine responses of various immune cells. Mice injected with THC affects Th 1 helper cell (↓) cytokines such as IFN-γ and IL-12 as well as IL-12 receptor, whereas T-helper 2 cytokines such as IL-4 and IL-10 increased (↑). Also, THC injection into mice increased catalepsy and shock along with serum IL-1, TNF-α, and IL-6. Human cell lines (major subpopulations) modulated by THC treatment have actual cytokine and chemokine production in culture. Human lung alveolar macrophages from marijuana smokers are deficient in functions such as phagocytosis, killing of bacteria, and suppressed production of TNF-α, GM-CSF, and IL-6 (with permission from ref. [7]).

on murine lymphocyte cultures also occur with human peripheral blood lymphocytes tested *in vitro* in similar types of studies. Thus, it is widely acknowledged that the detrimental effects of THC on resistance to *L. pneumophila* infection are related to modulation of lymphocyte and macrophage responses, as well as the cellular factors they produce such as IFN-γ and cytokines important in enhancing host resistance.

3. LEGIONELLA INFECTION AND THC

The Legionella infection model provided novel information concerning the effects of THC on both primary and secondary immunity.[2,5–7,17–19,22,29,30] BALB/c inbred mice are relatively resistant to infection by this organism but A/J mice are relatively susceptible due to specific genetic resistance factors. Experiments showed that BALB/c mice survived a primary infection with Legionella even when treated with THC either the day before or day after infection. In addition to surviving this primary infection, Legionella primed mice developed secondary immunity to subsequent infection with the same microbe administered several weeks later. Experiments were performed to assess the effects of THC on such secondary immunity. For this purpose, Legionella primed mice were challenged with a lethal dose of bacteria. Mice given only the primary sublethal infection developed secondary immunity and readily survived the secondary challenge infection. However, mortality after such challenge infection was significantly increased in mice treated with THC at the time of priming, indicating that injection of the cannabinoid at the time of primary infection suppressed development of expected secondary immunity.

Since marijuana is generally used more than once by individuals, experiments were performed to examine resistance to infection in mice treated with two doses of THC given 1 day before and 1 day after primary infection. No effects on survival were observed, although morbidity (e.g., malaise, etc.), was greater in the animals given only one injection. However, by increasing the drug dose, mortality increased dramatically beginning as soon as 30 min following the second THC injection. These results indicated that THC injection into mice significantly modified the course of both primary and secondary infection with Legionella.

Legionella infection induces mobilization of cytokines. Toxic shock-like death following THC treatment and primary infection indicated that administration of THC coincidental with infection mobilized cytokines to toxic levels.[17,29] It is widely accepted that this shock is due to acute phase cytokines and indeed it was found that levels of these cytokines (e.g., TNF-α and IL-6) increased markedly in the serum of the Legionella infected and THC injected mice. Furthermore, administration of monoclonal antibodies to either TNF-α, IL-6, or IL-1 protected the mice from this drug-induced mortality.[17] Changes in arachidonic acid metabolites appeared involved because production of these metabolites and cytokines are known to be closely linked and THC treatment is associated with changes in arachidonic acid metabolism.

4. THC EFFECTS ON T-CELL BIASING

We reported that THC injection into mice suppressed Th1 immunity by inhibiting mobilization of IFN-γ and IL-12 as well as the expression of IL-12 receptors, and increased Th2 immunity by promoting the cytokine IL-4.[30] Other investigators have reported similar T-helper cell biasing effects of cannabinoids, as well as other neuroimmune-axis modulating agents. For example, THC injection into mice enhanced development of a tumor enhancing Th2 helper cell cytokine response, and THC treatment of cultured human peripheral blood cells shifted responses to Th2 immunity.[31,32] These effects appeared mediated by activation of cannabinoid receptors.

It is now well accepted that there are two major receptors for cannabinoids, that is, CB1 mainly in the brain and CB2 mainly on immune cells.[32,33] A number of specific cannabinoid receptor antagonists have been developed, including a receptor antagonist for CB1 and a separate one for CB2. Studies in our laboratory using these receptor antagonists indicated that THC suppressed Th1 biasing activity, including IL-12 receptor activity, by a CB1-mediated mechanism (Fig. 3) and enhanced Th2 helper cell biasing by a CB2 mechanism, affecting GATA 3 (Fig. 4), a transcription factor which promotes Th2 cell differentiation.[35,36] Biasing toward Th2 immunity suppresses resistance to Legionella infection.

Beside macrophages, dendritic cells are an extremely important cell type for innate immunity involved in host resistance to opportunistic microbes, including bacteria like Legionella.[35] Myeloid cells differentiate into macrophages and further mature into dendritic cells. Lymphoid cells and dendritic cells produce either IL-4 or IL-12 in response to antigen stimulation, and production of these

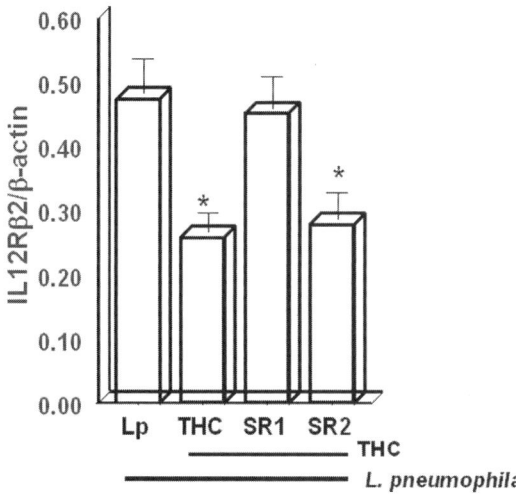

FIGURE 3. THC suppresses IL-12Rβ2 mRNA by CB1-mediated mechanism. BALB/c mice injected i.v. with either saline or THC (8 mg/kg) 18 hr prior to infection with *Legionella pneumophila* (7×10^6 bacteria). Two hours after infection, spleens were removed, total RNA extracted and analyzed by semi-quantitative RT-PCR for IL-12 receptor and β-actin mRNA. CB1 or CB2 antagonists injected 30 min prior to THC treatment (with permission from ref. [5]).

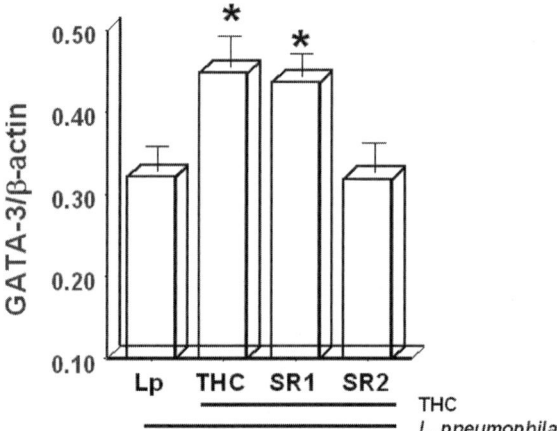

FIGURE 4. THC injection increases GATA 3 mRNA by CB2-mediated mechanism. BALB/c mice injected i.v. with either saline or THC (8 mg/kg) 18 hr prior to infection with *L. pneumophila* (7×10^6 bacteria), and 2 hr later, spleen was removed, total RNA extracted, and RT-PCR for GATA 3 and bactin mRNA determined. CB1 or CB2 antagonists injected 30 min before THC (with permission from ref. [5]).

factors is one of the many key steps in the development of Th1 vs Th2 cells[35,36] (Fig. 1). IL-12 leads to phosphorylation of STAT 4 and subsequent activation of IFN-γ and IL-12 receptor genes. The upregulation of these genes drives development of Th1 cells. On the other hand, the cytokine IL-4 leads to phosphorylation of STAT 6 and subsequent activation of genes for the IL-4 receptor promotes Th2 cell development. In the relative absence of IL-12, IL-4 stimulation increases the Th2-dependent transcription factor GATA 3, further promoting Th2 cell differentiation. Studies in this laboratory showed that THC markedly affected the regulation of expression of these key T-helper cell biasing mediators, IL-12 receptor and GATA 3 (Figs 3 and 4).

Mice injected with THC 18 hr prior to infection with Legionella develop increased susceptibility to these bacteria. In experiments to examine mechanisms involved, it was found that 2 hr following Legionella infection of THC-treated mice, spleen cells from the animals processed for measurement of IL-12 receptor mRNA by RT-PCR revealed that, like IL-12, THC treatment suppressed activation of this gene. Mice injected with either the CB1 or CB2 antagonist 30 min prior to THC injection showed that the CB1, but not CB2 antagonist, markedly attenuated the suppressive effect of the cannabinoid on Legionella resistance, indicating that downregulation of IL-12 receptor gene was indeed mediated by CB1.

Since reports from this laboratory have shown that THC alters the increased production of IL-4 by spleen cells from Legionella-infected mice, it seemed likely that GATA 3 could be important in this response. This signaling factor is important in regulation of T-helper cell development of Th2 but not Th1 cells. Furthermore, GATA 3 and IL-12 have a mutual antagonistic interaction, with the former increasing activity in the absence of the latter. Because THC suppressed

IL-12 production by macrophages or dendritic cells, experiments were performed to determine whether this cannabinoid affected GATA 3 production following Legionella challenge. For this purpose, spleens were harvested from mice 3 hr following Legionella infection and total RNA for GATA 3 mRNA determined by RT-PCR. It was found that GATA 3 message was indeed increased because of THC injection, indicating that this Th2 biasing transcription factor increased due to the drug treatment and was related to altered susceptibility to Legionella infection. To determine if cannabinoid receptors were involved in this effect, mice were pretreated with either CB1 or CB2 antagonists before THC injection and Legionella challenge. The CB2 antagonist but not the CB1 antagonist attenuated the THC effect, indicating that this marijuana cannabinoid increased GATA 3 gene activity through CB2 but not CB1.

5. DISCUSSION AND CONCLUSIONS

Many studies with the widely used illegal drug of abuse, marijuana, have shown that this drug affects host susceptibility to microbial infection. In this regard, previous studies have shown that the major component of marijuana, the cannabinoid THC, markedly alters susceptibility of mice to challenge infection with · L. pneumophila, an important and ubiquitous opportunistic bacterial pathogen which causes about 25,000 cases of Legionnaires disease per year in the Unites States alone. Injection of mice with THC prior to challenge infection with these bacteria suppresses the cytokines IL-12 and IFN-γ, considered important indicators of Th1 helper cell activity. This suppressive effect of THC was attenuated by antagonists to both the CB1 and CB2 cannabinoid receptors. The effects of CB1 and CB2 involvement were found split between suppression of IL-12 receptor gene and an increase in GATA 3 message. Unlike effects on IL-12 and IFN-γ production, THC suppression of IL-12 receptor was mediated only by CB1 whereas the THC-induced increase in GATA 3 was found mediated by CB2 only.

It is widely known that cannabinoid receptors are typical G-protein-coupled receptors, being coupled to Gi and suppressing adenylcyclase. Activation of G proteins through receptor ligation sets into motion a cascade of signaling and gene activation events mediated not only by the Gα component but also by the β/γ component of the G protein. Various signaling factors are activated or suppressed by G-protein activation, and the dominant pathways vary from cell to cell, depending upon endogenous receptors, the complement of neighboring receptors in the membrane and other unknown factors. Because CB1 mediated a decrease in IL-12 receptor gene activity and an increase in CB2-mediated GATA 3 gene, it appears that differences in regulation of these genes by cannabinoid receptors may be due to differences in the receptor and signaling milieu in Th1 vs Th2 cells. In other words, CB1 receptors on Th1 cells and the activated Gi subunits may signal a decrease in the IL-12 receptor gene product and, in contrast, CB2 receptor ligation on Th2 cells may signal an increase in the GATA 3 gene. Thus, it seems apparent that the cannabinoid system significantly impacts the function of the cytokine network in the immune system and this association is important for understanding mechanisms of host immunity to important opportunistic pathogens like Legionella. Further analysis of mechanisms concerning cannabinoid effects on host resistance to opportunistic bacteria, studied *in vivo* in animal models or *in vitro* with immune cell and humoral factors derived from infected animals, are warranted.

REFERENCES

1. Blanchard, D. K., Friedman, H., Stewart, II, W. E., Klein, T. W., and Djeu, J. Y. (1988). Role of γ-interferon in induction of natural killer activity by *Legionella pneumophila in vitro* and in an experimental murine infection model. *Infect. Immun.* **56:**1187–1193.
2. Friedman, H., Widen, R., Klein, T., and Johnson, W. (1984). Immunostimulation by *Legionella pneumophila* antigen preparations *in vivo* and *in vitro. Infect. Immun.* **43:**347–352.
3. Yamamoto, Y., Klein, T. W., Newton, C. A., and Friedman, H. (1992). Differing macrophage and lymphocyte roles in resistance to *Legionella pneumophila* infection. *J. Immunol.* **148:**584–589.
4. Blanchard, D. K., Newton, C., Klein, T. W., Stewart II, W. E., and Friedman, H. (1986). *In vitro* and *in vivo* suppressive effects of Δ^9-tetrahydrocannabinol on interferon production by murine spleen cells. *Intl. J. Immunopharmacol.* **8:**819–824.
5. Klein, T. W., Newton, C., Larsen, K., Chou, J., Perkins, I., Lu, L. *et al.* (1993). Cannabinoid receptors and T helper cells. *J. Neuroimmunol.* **147:**91–94. 2004.
6. Newton, C. A., Klein, T. W., and Friedman, H. (1994). Secondary immunity to *Legionella pneumophila* and Th1 activity are suppressed by delta-9-tetrahydrocannabinol injection. *Infect. Immun.* **62:**4015–4020.
7. Klein, T. W., Lane, B., Newton, C. A., and Friedman, H. (2000). The cannabinoid system and cytokine network. *Proc. Soc. Exp. Biol. Med.* **225:**1–8.
8. Bradley, S. G., Munston, A. E., Dewey, W. L., and Harris, L. S. (1977). Enhanced susceptibility of mice to combinations of Δ^9-tetrahydrocannabinol and live or killed gram-negative bacteria. *Infect. Immun.* **17:**325–329.
9. Cabral, G. A., Lockmuller, J. C., and Mishkin, E. M. (1986). Δ^9-Tetrahydrocannabinol decreases α/β-interferon response to herpes simplex virus type 2 in the B6C3F1 mouse. *Proc. Soc. Exp. Biol. Med.* **181:**305–311.
10. Cabral, G. A., Mishkin, E. M., Marciano-Cabral, F., Coleman, P., Harris, L., and Munson, A. E. (1986). Effect of Δ^9-tetrahydrocannabinol on herpes simplex virus type 2 vaginal infection in the guinea pig. *Proc. Soc. Exp. Biol. Med.* **182:**181–186.
11. Cabral, G. and Dove Pettit, D. (1998). Drugs and immunity: Cannabinoids and their role in decreased resistance to infectious diseases. *J. Neuroimmunol.* **83:**116–123.
12. Juel-Jensen, B. F. (1972). Cannabis and recurrent herpes simplex. *Br. Med. J.* **4:**296.
13. Morahan, P. S., Klykken, P. C., Smith, S. H., Harris, L. S., and Munson, A. E. (1979). Effects of cannabinoids on host resistance to *Listeria monocytogenes* and herpes simplex virus. *Infect. Immun.* **23:**670–674.
14. Specter, S., Lancz, G., Westrich, G., and Friedman, H. (1991). Delta-9-tetrahydrocannabinol augments murine retroviral induced immunosuppression and infection. *Int. J. Immunopharmacol.* **13:**411–417.
15. Friedman, H., Klein, T., and Specter, S. (1991). Immunosuppression by marijuana and components. In R. Ader, D. L. Felten, and H. Cohen (eds), *Psychoneuroimmunology,* 2nd edn. Academic Press, New York, pp. 931–953.
16. Friedman, H., Shivers, S., and Klein, T. W. (1994). Drugs of abuse and the immune system. In J. Dean, M. Luster, A. Munson, and I. Kimber (eds), *Immunotoxicology Immunopharmacology,* 2nd edn. Raven Press, New York, pp. 303–322.
17. Klein, T. W., Newton, C., Widen, R., and Friedman, H. (1993). Δ^9-tetrahydrocannabinol injection induces cytokine-mediated mortality of mice infected with *Legionella pneumophila. J. Pharmacol. Exp. Ther.* **267:**635–640.

18. Klein, T. W., Newton, C., and Friedman, H. (1994). Resistance to *Legionella pneumophila* suppressed by the marijuana component, tetrahydrocannabinol. *J. Infect. Dis.* **169**:1177–1179.

19. Klein, T. W., Newton, C. A. and Friedman, H. (2001). Cannabinoid and the immune system. *Pain Res. Manage.* **6**:95–101.

20. Huber, G. L., Simmons, G. A., McCarthy, C. R., Cutting, M. B., Laguarda, R., and Pereira, W. (1975). Depressant effect of marijuana smoke on antibacterial activity of pulmonary alveolar macrophages. *Chest* **68**:769–773.

21. Mann, P. E. G., Cohen, A. B., Finley, T. N., and Ladman, A. L. (1971). Alveolar macrophages: Structural and functional differences between nonsmokers and smokers of marijuana and tobacco. *Lab. Invest.* **25**:111–120.

22. Klein, T., Friedman, H., and Specter, S. (1998). Marijuana immunity and infection. *J. Neuroimmunol.* **83**:102–115.

23. Lopez-Cepero, M., Friedman, M., Klein, T., and Friedman, H. (1986). Tetrahydrocannabinol-induced suppression of macrophage spreading and phagocytic activity *in vitro. J. Leukoc. Biol.* **39**:679–686.

24. Rachelefsky, G. S., Opelz, G., Mickey, M. R., Lessin, P., Kiuch, M., Silverstein, M. J. *et al.* (1976). Humoral and cell-mediated immunity in chronic marijuana smoking. *J. Allergy Clin. Immunol.* **58**:483–490.

25. Raz, A. and Goldman, R. (1976). Effect of hashish compounds on mouse peritoneal macrophages. *Lab. Invest.* **34**:69–76.

26. Zimmerman, S., Zimmerman, A., M., Cameron, I. L., and Laurence, H. L. (1977). Δ^1-Tetrahydrocannabinol, cannabidiol, and cannabinol effects on immune response of mice. *Pharmacology.* **15**:10–23.

27. Specter, S., Lancz, G., and Friedman, H. (1990). Marijuana and immunosuppression in man. In R.R. Watson (ed.), *Drugs of Abuse and Immune Function*, CRC Press, Boca Raton, FL, pp. 73–85.

28. Pross, S., Klein, T., Newton, C., and Friedman, H. (1987). Differential effects of marijuana components on proliferation of spleen, lymph node and thymus cells. *Int. J. Immunopharmacol.* **9**:363–370.

29. Klein, T., Newton, C., and Friedman, H. (1998). Cannabinoid receptors and immunity. *Immunol. Today* **19**:373–381.

30. Klein, T. W., Newton, C. A., Nakachi, H., and Friedman, H. (2000). Δ^9-Tetrahydrocannabinol treatment suppresses immunity and early IFN-γ, IL-12, and IL-12 receptor $\beta2$ responses to *Legionella pneumophila* infection. *J. Immunol.* **164**:6461–6466.

31. Yuan, M., Kiertscher, S. M., Cheng, Q., Zoumalan, R., Tashkin, D. P., and Roth, M. D. (2002). Δ^9-Tetrahydrocannabinol regulates Th1/Th2 cytokine balance in activated human T cells. *J. Neuroimmunol.* **133**:124–131.

32. Zhu, L. X., Sharma, S., Stolina, M., Gardner, B., Roth, M. D., Tashkin, D. P. *et al.* (2000). Δ^9-Tetrahydrocannabinol inhibits antitumor immunity by a CB2 receptor-mediated, cytokine-dependent pathway. *J. Immunol.* **165**:373–380.

33. Munro, S., Thomas, K. L., and Abu-Shaar, M. (1993). Molecular characterization of a peripheral receptor for cannabinoids. *Nature* **365**:61–65.

34. Shire, D., Calandra, B., Rinaldi-Carmona, M., Oustric, D., Pesegue, B., Bonnin-Cabanne, O. *et al.* (1996). Molecular cloning, expression, and function of the murine CB2 peripheral cannabinoid receptor.*Biochim. Biophys. Acta* **1307**:132–136.

35. Moser, M. and Murphy, K. (2000). Dendritic cell regulation of Th1–Th2 development. *Nat. Immunol.* **1**:199–205.

36. Ouyang, W., Ranganath, S. H., Weindel, K., Bhattacharya, D., Murphy, T. L., Sha, W. C. *et al.* (1998). Inhibition of Th1 development mediated by GATA 3 through an IL-4-independent mechanism. *Immunity* **9**:745–755.

6

Nicotine and Immunity

SUSAN PROSS and HERMAN FRIEDMAN

1. INTRODUCTION

Nicotine, a small organic alkaloid synthesized by tobacco plants, is the addictive component of cigarettes.[1] Its basic properties permit easy transport across the small intestine and lung tissues into the blood. Nicotine's size and lipophilic characteristics allow for a small amount to cross cell membranes directly, without interception by a receptor,[1] although its primary effects are via receptor mediation. This small alkaloid acts as an agonist at the nicotinic acetylcholine receptors (nAChRs), found mainly in the central (CNS) and peripheral nervous system, as well as on many other tissue cells throughout the body.[2] The distribution of these receptors on a large variety of cells helps to explain why nicotine has been associated with a wide range of biological actions. These actions of nicotine account, in part, for alterations in the cardiovascular, pulmonary, gastrointestinal, urogenital, hepatic, and nervous systems caused by smoking tobacco.[3–7] The most frequent way to acquire nicotine is via tobacco products. Cigarettes contain approximately 1.5–2.5 mg nicotine per cigarette, with the highest level of nicotine reported in the plasma of heavy smokers being about 700 ng/ml.[8,9] Interestingly, different areas of the body accumulate nicotine at different rates. For example, nicotine is retained at a higher level in the cervix,[10] kidneys, gastrointestinal tract, heart, and muscles[11] than in blood. In terms of distribution in the blood, smoking one cigarette results in about 50 ng nicotine/ml in arterial blood contrasted with the 20 ng nicotine/ml in venous blood.[12] This type of differential distribution of nicotine is important to keep in mind when comparing results of nicotinic action from different experimental protocols.

SUSAN PROSS and HERMAN FRIEDMAN • Department of Medical Microbiology and Immunology, University of South Florida College of Medicine, Tampa, FL 33612.

Infectious Diseases and Substance Abuse, edited by Herman Friedman *et al.*
Springer, New York, 2005.

It is nicotine, just one of the thousands of components of tobacco, that is most strongly related to the addictive consequences of smoking. This addictive characteristic is best explained by the intermittent acquisition of nicotine, which can travel within 8 s to the brain, and the subsequent intermittent release of dopamine in the brain.[13] Importantly, chewing tobacco, or smokeless tobacco, also delivers a similar amount of nicotine to the blood; however, the distribution is slower and the timing continuous.[14,15] Since high concentrations of nicotine have been reported in the saliva of snuff users (up to 5 μg/ml), it has been suggested that nicotine may be important in the induction of oral cancers in people using smokeless tobacco.[16]

At one time, the use of tobacco products was the most direct way of being exposed to nicotine. However, since it is becoming more common for people to use nicotine gums, inhalers, or patches as substitutes for tobacco, nicotine can therefore be acquired independent of the other chemicals in cigarette smoke or chewing tobacco. This use of nicotine is potentially important since nicotine engages many organ systems. It is well supported that nicotine's action *in vivo* impinges on the hypothalamic–pituitary–adrenal (HPA) axis,[17] and thus its effects are broadbased throughout the body. Specifically, it is known that nicotine enhances the release of neurotransmitters and hormones, including acetylcholine, serotonin, dopamine, norepinephrine, prolactin, vasopressin, and corticosteroids.[18] These components have their own modulatory actions on the body, extending the potential impact of nicotine. Further, nicotine has been reported to act both via receptor and nonreceptor-mediated mechanisms, again extending the range of its potential effects.

This broad action of nicotine is important to consider since it is used by such a large number of people. In this regard, in addition to its use in facilitating the stopping of smoking, nicotine has been given to people with a variety of diseases since it has been shown to have some measurable clinical benefits.[13] For example, in the CNS, nicotine can increase short-term attention, cognition and memory, increase brain energy metabolism, and decrease hunger resulting in decreased body weight. It is used with some success to treat Alzheimer's disease,[19] to enhance cognitive function, to facilitate dopamine release from neurons thus relieving symptoms of Parkinson's disease, to reduce the severity of involuntary tics in Tourette's syndrome, and to aid patients with inflammatory bowel disease or attention deficit disorder.

There are many reports demonstrating modulation of immune parameters by nicotine in laboratory experiments; however, there have been no confirming data about the long-term effects of nicotine on immunity in clinical cases. Although the mechanisms of action of nicotine in immune cells are still unclear, data suggest that binding to the nAChR brings about changes in intracellular calcium levels, resulting in alterations of cell signaling pathways. These alterations would then be expected to promote modulations in immune cell activity such as increasing cytotoxicity[20] and inducing T-cell anergy *in vivo*.[21,22] Although much of the literature supports that the mechanism of action of nicotine is through the nAChR, some reports suggest that the mechanism of action of nicotine may in some cases be independent of the nAChR. In fact, recent work reported that nicotine contributed to neutrophil accumulation in smoke-associated lung diseases by enhancing the survival of these cells, and that the mechanism of action of nicotine was through noncholinergic receptor binding, without activation of protein kinases.[23] The interaction of nicotine and the nAChR is described below.

2. NICOTINIC ACETYLCHOLINE RECEPTORS AND GENERAL PHYSIOLOGICAL EFFECTS

The nAChRs are pentameric transmembrane ion channels that open when acetylcholine or its agonists are bound, allowing Na^+ or Ca^{2+} ions to cross into the cell, activating second messenger signaling pathways that result in *de novo* protein synthesis that change cell function or activity. The channels are made of five of the same subunits (homopentamers) or arrangements of different subunits (heteropentamers)—usually, 2 alpha and 3 beta subunits are found in neuronal nAChRs. At the neuromuscular junction, 2 alpha, 1 beta, 1 delta, and 1 gamma or 1 epsilon subunits are involved in the receptor structure. Acetylcholine or nicotine binds the alpha subunits with assistance from the beta subunits. Presently, 10 alpha subunits (alpha 1–10) and 4 beta subunits (beta 1–4) have been demonstrated. There are two broad categories of acetylcholine receptors—muscarinic (found in the CNS, autonomic ganglia, and parasympathetic effector cells), and nicotinic (found in the CNS, neuromuscular junctions, and autonomic ganglia).[24] Nicotinic acetylcholine receptors are found on nonneuronal cells and the existence of nAChRs on immune cells has been demonstrated by pharmacological studies.[2,25,26] The endogenous ligand for nAChRs is acetylcholine, and nicotine acts as an agonist when bound. Muscarinic acetylcholine receptors also claim acetylcholine as their endogenous ligand; however, muscarine acts as an agonist at muscarinic AChRs where nicotine has no effect, and nicotine acts as an agonist at nicotinic AChRs where muscarine has no effect.

In the CNS, nicotine causes the release of neurotransmitters and hormones including acetylcholine, dopamine, serotonin, ACTH, beta-endorphin, prolactin, epinephrine, and norepinephrine.[13] In the efferent peripheral nervous system, nAChRs are found at the neuromuscular junction and at the sympathetic and parasympathetic postsynaptic ganglia. At the sympathetic and parasympathetic ganglia, use of nicotine can result in increased heart rate, constricted blood vessels, decreased skin temperature, increased gastrointestinal activity, increased circulating fatty acids, and increased secretion of epinephrine and norepinephrine from the adrenal gland resulting in general systemic stimulation.[1,13] Nicotine alters lipid metabolism by increasing circulating total cholesterol, phospholipids, triglycerides, very low-density lipoproteins (LDLs) and low-density lipoproteins.[27] These metabolic changes, coupled with nicotine-induced alterations in the integrity of blood vessels,[14] support a role of nicotine in cardiovascular disease.[27]

3. NICOTINE AND IMMUNITY

3.1. General

Recent research on nicotine suggests that it can affect the normal physiology of various tissues, including those of the immune system. The nicotinic acetylcholine receptor protein has been found on both intact lymphocytes and lymphocyte membranes,[28,29] and the mRNA of the alpha 2–7 and beta 2–4 subunits of the nicotinic acetylcholine receptor has been found in human peripheral blood mononuclear cells.[2,30] However, the functional role of nicotinic acetylcholine receptors in nicotine-induced immunomodulation has not been clarified. The discoveries of nicotinic acetylcholine receptors on immune cells give mechanistic support to the hypothesis that nicotine alters immune cell functions via its receptor. A typical step in the signal transduction pathway of nicotine is altered intracellular calcium concentrations. Nicotine exposure resulted in a downregulation of intracellular calcium after immunostimulation of T- and B cells when these cells were exposed to nicotine *in vivo*[31] and an upregulation of intracellular calcium after immunostimulation of human peripheral blood cells or leukemic cell lines when these cells were exposed to nicotine *in vitro*.[32] These data demonstrated an effect of nicotine on T-cell signal transduction even though opposing results were obtained with different experimental protocols.

It has been reported by us and by others that nicotine impacts the production of various cytokines, indicating action of nicotine on specific immune cells.[33–35] Figure 1 shows that nicotine can decrease the production of TNF-α induced by LPS stimulation of adherent murine splenocytes. TNF-α is cytokine produced by many cells including macrophages and is an example of an inflammatory cytokine. Results such as this support the concept that nicotine is anti-inflammatory.

Depending upon the types of cytokines they produce, specific immune cells can be categorized into Th2 cells producing IL-10, Th1 cells producing IFN-γ, or inflammatory cells producing IL-6 and TNF-α, among other cytokines and chemokines. Interestingly, when transdermal nicotine is given to healthy, male nonsmokers, peripheral blood mononuclear cells from these volunteers produced less IL-10, while the amount of IFN-γ produced was unchanged.[36]

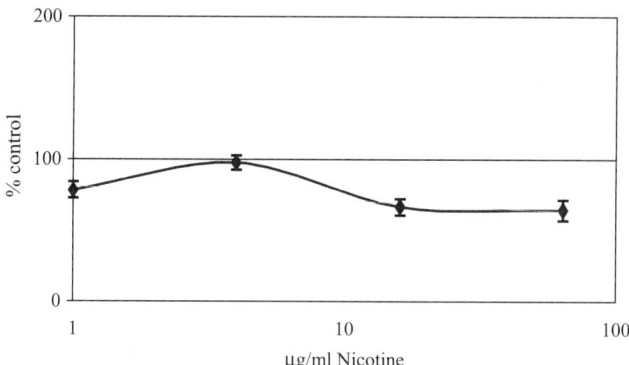

FIGURE 1. Effect of different nicotine concentrations on TNF-α production by LPS-stimulated adherent murine splenocytes. TNF-α production was measured by ELISA and calculated as a percentage of unexposed (no nicotine) controls. Kruskal-Wallis one-way ANOVA and Student Newman Keuls multiple comparisons tests showed that higher concentrations (16 and 64 μg/ml) of nicotine significantly inhibited TNF-α production as compared to adherent splenocytes stimulated with LPS and not exposed to nicotine. Data are presented as means $\pm$ SEM within nicotine concentration; $n = 5$–6, $p = 0.001$.

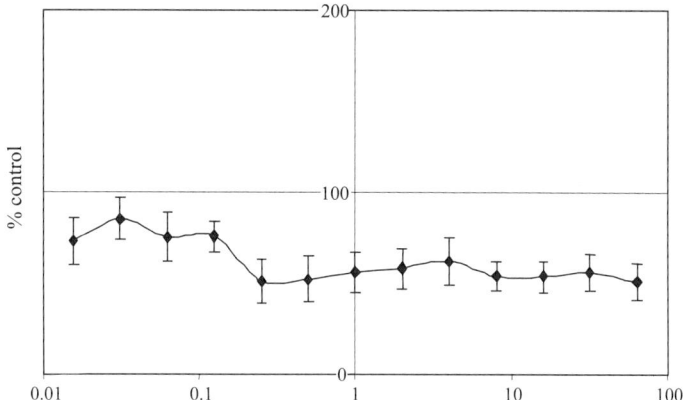

FIGURE 2. Effect of nicotine on IL-10 production by murine splenocytes. Splenocytes from mice were immunostimulated with ConA and concurrently exposed to nicotine for 48 hr. IL-10 production was measured by ELISA and calculated as a percentage of unexposed (no nicotine) controls. Kruskal-Wallis one-way analysis of variance followed by Student Newman Keuls multiple comparisons test showed that nicotine significantly decreased IL-10 production by splenocytes from young adult mice at the higher concentrations of nicotine tested (0.25–2 and 8–64): $p = 0.003$, $n = 5$–10.

This downregulation of nicotine on IL-10, with no effect on IFN-γ has also been reported by our laboratory.[35] See Fig. 2.

Specifically, when nonadherent murine splenocytes were exposed to concanavalin A and nicotine *in vitro*, the production of IL-10 was diminished while the production of IFN-γ was unchanged. Thus, nicotine, independently from smoking, inhibited this Th2 immune cell function, while presumably allowing the Th1 response to occur. Nicotine-induced modulation of cytokine production was achieved at 1 μg/ml, which is within the range of nicotine levels measured in the serum of smokers.[12] When looking at inflammatory cytokines, nicotine exposure resulted in either decreased or no change in the production of the cytokines studied. In our studies, nicotine significantly inhibited TNF-α production by LPS-stimulated adherent murine splenocytes,[37] similar to findings by other researchers that TNF-α production by human peripheral blood mononuclear cells was inhibited by nicotine *in vitro*.[38] In contrast, IL-6 production by LPS-stimulated adherent murine splenocytes was not affected by nicotine exposure,[37] similar to results found in smokers.[39] However, the data obtained from smokers may not be comparable to the data obtained from exposure to nicotine alone due to the thousands of other chemicals found in cigarette smoke. Other immune cell functions altered by nicotine include increased cytotoxicity[20] and induced T-cell anergy *in vivo*.[21,22]

Most reports describe the action of nicotine as being due to its interaction with the nAChR. However, some reports suggest that the mechanism of action of nicotine is not always through the nAChR. In fact, researchers have reported that nicotine contributed to neutrophil accumulation in smoking-associated lung diseases by enhancing the survival of these cells. Furthermore, the mechanism of action was found to be through noncholinergic receptor binding, without activation of protein kinases.[23] Thus, nicotine does affect immune cells and immune responses in various ways. However, the link between the long-term effects of nicotine on immune responses and the clinical wellness of individuals has not been firmly established.

Apoptosis, or programmed cell death, is a normal physiologic process by which the body removes unwanted cells. This process is critical in embryonic development as well as in tissue homeostasis in the adult. Apoptosis results in a nonreversible removal of cells and is thus of paramount importance in maintaining homeostasis in a system where cells are continually being generated, such as the immune system.[40,41] Inhibition of apoptosis in antigen-stimulated immune cells can result in difficulties in mounting a full defense against viral or bacterial infection by decreasing the proportion of high-affinity antigen-specific immune cells. Inhibition of apoptosis in precancerous cells can result in the promotion of tumor growth by allowing mutated cells to survive. Two specific types of cell death have been described, necrosis and apoptosis,[42] which differ significantly in terms of their morphologic and biochemical characteristics. Necrosis, a process that is usually initiated by cell injury, mechanical or chemical, is characterized by cellular swelling and inflammatory response. Apoptosis, unlike necrosis, is an active process, which is genetically programmed. The induction of apoptosis depends on soluble mediators, cell-to-cell contacts, intracellular signaling, transcription factor activation, or cytoplasmic second messengers. This self-destruction process involves the activation of a family of cysteine proteases (caspases) that play a key role in apoptosis. The caspases are present as inactive proenzymes that appear to be constitutively expressed in most cells. Activation of these caspases allows the execution of the effector phase of cell death.[43,44] Hence in mammals, this core effector mechanism of programmed cell death is regulated upstream by signals involved in cellular differentiation and cellular proliferation, allowing the whole organism to control the fate of each of its cells in a refined and complex manner. Multiple agents, including glucocorticoid hormones, induce apoptotic-signaling pathways.[45] Glucocorticoids affect a variety of tissues and body systems, and their role in the immune system is central for induction of cell cycle arrest and the programmed cell death of both immature thymocytes and peripheral T lymphocytes.[46] Dexamethasone (DEX), a synthetic glucocorticoid hormone, induces apoptosis through binding to the glucocorticoid receptors. These receptors are found in an inactive state within the cytoplasm and become activated when bound. Activation of these receptors leads to cascading events that include the production of active caspases, the repression of genes necessary for cell proliferation, and the transcriptional upregulation of responsive lysis genes.[47]

Morphologically, apoptosis is characterized by cytoplasmic condensation, and intranucleosomal cleavage of DNA by endonucleases present within the cell, and DNA fragmentation into 180–200 base pairs (BP) of the dying cell itself, all with minimal inflammation.[48] Faulty regulation of apoptosis has been implicated in degenerative conditions, vascular diseases, AIDS, and cancer.[49] Evidence shows that uncontrolled induction of apoptosis may lead to diseases as diverse as Alzheimer's and Hodgkin's diseases, as well as to exacerbate the course of autoimmunity in general.[50,51] Conversely, apoptosis can have a very positive effect in terms of protection against diseases by facilitating death in tissues whose growth is out of control. For example, one mechanism whereby tumor cells gain immortality is by loss of their ability to undergo apoptosis.[52,53] Thus, the acquisition of resistance to apoptosis would confer a survival advantage to emerging tumor cells.[48,54]

In terms of the association of nicotine and apoptosis, it has been hypothesized that nicotine abuse could decrease survival of progenitor populations in the developing and adult brain through this process. Similarly, nicotine has been shown to have a dose-dependent ability to induce cytotoxicity in human glioma and glioblastoma cell lines.[20] In contrast, the action of nicotine on apoptosis seems to differ in neutrophils, whereas it has been shown that nicotine suppresses apoptosis.[23] Studies in our laboratory (see Fig. 3) have shown that nicotine inhibited the expression of caspase-3 in immune cells treated with dexamethasone (DEX). Treating the cells with d-tubocurarine chloride, an antagonist at nicotinic receptors, blocked this inhibition. Since thymocytes need to undergo apoptosis for proper selection within the thymus, and splenocytes need to undergo apoptosis to maintain appropriate handling of foreign antigens, the role of nicotine as an inhibitor of the process of apoptosis is significant. Furthermore, the inducer of apoptosis in these studies was DEX, a synthetic glucocorticoid. Glucocorticoid production in the body is highly dependent on environmental challenges, such as stress, which also have an impact on immunity. The combined action of nicotine with peripheral as well as CNS involvement is significant and warrants further investigation.

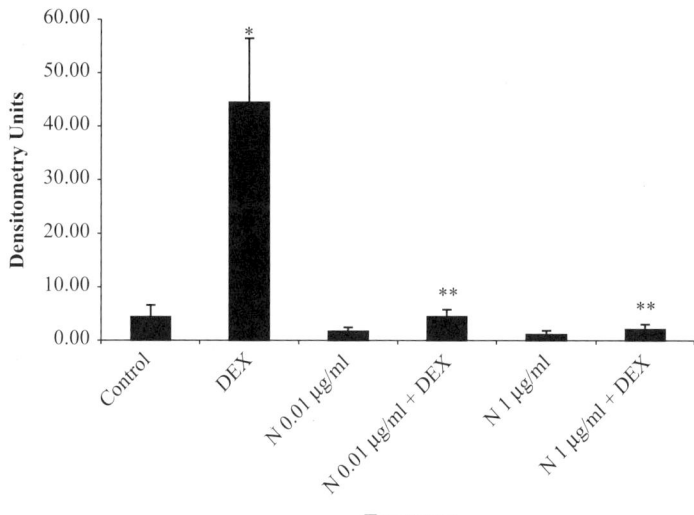

FIGURE 3. Effect of nicotine on murine splenocytes. Murine splenocytes were incubated for 3 hr in either of 6 groups: untreated cells (Control); cells with 100 nM dexamethasone (DEX); cells with nicotine alone (Nic 0.01 µg/ml or Nic 1 µg/ml); and cells with nicotine and DEX (0.01 µg/ml nicotine + DEX, or 1 µg/ml nicotine and DEX). *, significantly different ($p < 0.05$) compared to the Control; **, significantly different ($p < 0.05$) compared to DEX.

Note that either nicotine concentration significantly decreased the expression of active caspase-3 compared to the DEX group. The DEX group showed a significantly enhanced expression of active caspase-3 compared with the Control and Nic 0.01 μg/ml and Nic 1 μg/ml groups ($p < 0.05$). The DEX-treated cells showed a significant decrease in expression of active caspase-3 when treated concurrently with nicotine.

4. RELATIONSHIP OF NICOTINE TO DISEASE

4.1. Nicotine and Cardiovascular Disease and Stroke

Over the past several decades, researchers have focused on the mechanisms involved in the initiation and progression of atherosclerosis and cardiovascular disease, with findings emphasizing the role of inflammation in response to endothelial injury induced by trauma and infection. The major events in the induction of atherosclerosis include vascular endothelial cell injury complicated by binding of monocytes to the vascular endothelium and subsequent inflammation, as well as the proliferation and migration of vascular smooth muscle cells.[55–57] In this regard, the effects of nicotine on infection and on inflammation may be critical. It is at this point, where the immune system intersects with the cardiovascular system, that nicotine may have a profound effect. Data have emphasized that nicotine was directly chemotactic for neutrophils and may have had adverse effects on their function.[58] Neutrophils are involved in the inflammatory response after trauma and are associated with an increased coronary vasoconstrictive response.[58] The actions of nicotine in accentuating cardiovascular disease have been gaining support due in part to its role in the oxidation of LDL.[27,59,60] Of additional importance is the ability of nicotine to alter the homeostatic profile of chemokines and cytokines that could promote the development of atherosclerosis and cardiovascular disease.[61]

Using a similar methodology as described in Fig. 1, it was found that human coronary artery endothelial cells (HCAECs) with no nicotine treatment (control) and with nicotine treatment (Nic 1 μg/ml) showed only a minimal level of expression of active caspases (Fig. 4).

Figure 4 shows that treatment of HCAECs with 100 nM DEX and 40 ng/ml TNF-α, the apoptosis inducers, resulted in approximately a 3-fold increase in the expression level of active caspases compared to the control levels. Co-treatment of these cell cultures with both 1 μg/ml of nicotine and the apoptosis inducers resulted in inhibition of apoptosis as evidenced by a decrease in the expression level of active caspases ($p < 0.05$), such that they were not significantly different from either those of the control cultures or cultures treated with nicotine alone. Thus, nicotine essentially prevented an increase in the expression level of active caspases by inhibiting the apoptosis process.

To determine whether the action of nicotine was receptor-mediated, 100 μM of d-TC was concurrently added to the cell cultures that were treated with both apoptosis inducers and nicotine. Blocking nicotinic receptors with

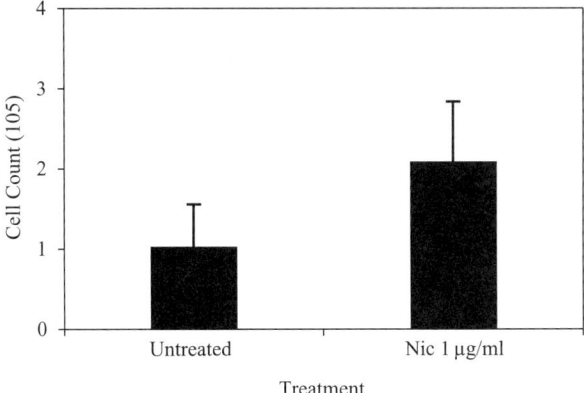

FIGURE 4. Role of nicotine on cell count of cultured HCAECs. The untreated group of cultured HCAECs (3.2×10^5/well) incubated for 48 hr at 37°C in a humid chamber under 5% CO_2 atmosphere showed a decrease of cell count compared to the zero time, while the group that was treated with nicotine (Nic 1 μg/ml) maintained the cell count ($p < 0.05$). Representative data is the mean of four experiments.

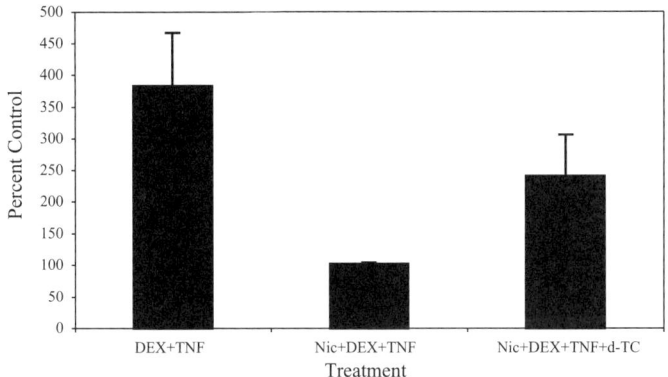

FIGURE 5. Induction of apoptosis in HCAECs with and without nicotine treatment. The blocking effect of d-TC was demonstrated by evaluating the expression level of active caspases in HCAECs. In all experimental groups, cultured HCAECs (1×10^6/ml), were aliquoted in 500 μl into 24 well plates and incubated for 3 hr at 37°C in a humid chamber under 5% CO_2 atmosphere. The nicotine concentration was 1 μg/ml, DEX at 100 nM, and TNF at 40 ng/ml. Representative data is the mean of 3-7 experiments and is expressed in percentage control.

d-TC (Fig. 5) significantly reversed the inhibitory effect of nicotine on apoptosis, as shown by the increased level of active caspases ($p < 0.05$). These levels were not significantly different from those cultures treated with apoptosis inducers (DEX + TNF) alone. There was no significant difference between the group treated with apoptosis inducers and the group of cells treated with the combination of nicotine, apoptosis inducers, and d-TC.

5. NICOTINE AND CANCER

In addition to nicotine being an addictive agent, it has been shown by some researchers to possess tumorigenic or tumor-promoting activities.[62,63] Research that has associated increased incidences of pancreatic cancer with smoking has highlighted the role of nicotine, proposing that nicotine levels can be correlated to cancer occurrence or progress.[64,65] The induction of pancreatic injury by nicotine may involve activation and expression of a proto-oncogene called ras. In terms of associating lung cancer with nicotine exposure, it has been noted that high affinity nAChRs are found on human lung cancer cells of many histological types as well as in normal lung tissue.[66–69] Nicotine has also been shown to enhance the growth of lung cancer cells *in vitro*, again suggesting a role in pulmonary carcinogenesis.[70,71] Furthermore, *in vitro* studies correlated nicotine exposure with the probability of cervical tumor progression in humans.[63,72,73] Recent studies have shown that nicotine, at concentrations found in smokers, can activate the serine/threonine kinase Akt in nonimmortalized human airway epithelial cells *in vitro* in a receptor-dependent manner, supporting the concept that nicotine could contribute to tobacco-related carcinogenesis.[74] Nicotine has been shown to stimulate colon cancer cell proliferation and tumor growth in a nude mouse xenograft model. In addition to stimulating SW1116 colon cancer cell proliferation in a dose-dependent manner, EGFR and c-Src phosphorylation levels as well as protein expression of 5-LOX were significantly enhanced. In addition, *in vivo* studies using a xenograft model showed that nicotine also significantly enhanced tumor growth. This acceleration of tumor growth corresponded well with increased vascularization and its pro-angiogenic factors.[75] Despite this relationship between nicotine and cancer, it needs to be recognized that not all researchers have demonstrated such a correlation.[76] For instance, nicotine has been shown to result in cytotoxicity in some cases, and prolongation of cell life in other cases.[23] Results are often dependent on cell type investigated, nicotine concentration, exposure time, and age of the individual.[20,77,78]

Interestingly, nicotine has been found to stimulate cytoplasmic and nuclear accumulation of growth factors such as transforming growth factor-beta (TGF-beta) as well as to inhibit lysosomal degradation of growth factors. This mechanism is considered a potential mechanism for tobacco-induced tumor promotion. Nicotine has been found to have an impact on intracellular cell-signaling pathways and activate the mitogen-activated protein kinase (MAPK) pathway.[79] Such actions of nicotine, often at concentrations of less than 1 μM, could lead to changes in cell growth by increasing apoptosis.[80] The studies supporting the notion that nicotine may have carcinogenic potential, either directly as a carcinogen or, more likely, indirectly as a promoter of cancer, often demonstrate that nicotine may enhance expression of oncogenes that block apoptosis.[81] For example, Yoshida *et al.* have hypothesized that, in order to induce the DNA fragmentation indicative of the apoptotic process, intracellular Ca^{2+} concentrations needed to rise above a threshold level and that nicotine might be interfering with this process in the cells tested. Several studies, including data from our laboratory,[82,83] have reported that nicotine blocks apoptosis in a wide variety of cell types.[84] This finding has been seen in protocols using a wide variety of stimuli in normal as well as in cancer cells, but the mechanism of action of nicotine, even as to whether the effect was receptor mediated, was not elucidated in many of these studies.[48] Other studies, focusing on human lung cancer cells, showed that nicotine blocked opioid-induced apoptosis by modulating intracellular signaling pathways involving PKC or MAPK.[66] Opioids such as morphine decreased PKC activity and increased apoptosis whereas nicotine increased total PKC activity and decreased apoptosis.[66,85–87] In contrast to the cited studies listed above, a recent investigation by Berger *et al.*[88] demonstrated that nicotine treatment enhanced apoptosis by inducing the expression of p53, a tumor suppressor protein. This cytotoxic action of nicotine was dependent on extracellular calcium levels, and cells that have difficulty buffering calcium, in this case immortalized hippocampal cells, were more susceptible to the cytotoxic action of nicotine. Interestingly, whereas some studies have demonstrated nicotine's interference with apoptosis and other studies have shown nicotine to be apoptotic on its own, still other research has shown that nicotine does not have any impact on apoptosis at all. The reasons for conflicting reports seem to relate to the tissues studied, animals investigated, concentrations of nicotine chosen, and time of exposure to nicotine.[20,66,81]

5.1. Nicotine and Infectious Disease

It is clear that tobacco smoking may predispose people to respiratory infection, but the data concerning alteration of response to infectious agents after exposure to nicotine alone is also becoming compelling. For example, Sopori *et al.* have shown that chronic exposure of mice and rats to nicotine diminishes T-cell activity as well as inflammation. Mice treated with nicotine and then exposed to the influenza virus have poorer outcomes than no nicotine-treated controls. The data support the concept that nicotine may impact negatively on infectious diseases that require inflammatory processes for protection.[22] In order to evaluate the action of nicotine on pulmonary activity, recent studies by Matsunaga *et al.* focused on the nicotine-induced modulation of antimicrobial activity and cytokine responses of alveolar macrophages to *Legionella pneumophila*, the etiological agent of Legionnaire's disease. The experimental model involved infecting MH-S alveolar macrophages with *L. pneumophila*, and then treating the cells with nicotine. Nicotine treatment of these macrophages downregulated the production of IL-6, IL-12, and TNF-α, but not IL-10. In addition to demonstrating an effect on cytokine action by nicotine, it was also found that this action of nicotine was receptor mediated, since the action was completely blocked by a nonselective antagonist, d-tubocurarine, for nAChRs.[89] This action of nicotine on bacteria was again shown in another bacterial model system. Yamaguchi *et al.* found that stimulation of nAChRs with nicotine altered the growth of *Chlamydia pneumoniae* in epithelial HEp-2 cells. This result is important because not only did it demonstrate a role for nicotine in infection, it also was more generalizable to a possible pathophysiological role of nAChRs in terms of intracellular infection.[90] The activity of nicotine in an *in vivo* model was clearly demonstrated in a recent experiment by Myles *et al.* in which nicotine alone, given to rabbits as a patch, could induce ocular shedding of herpes simplex virus type 1 (HSV-1) in rabbits that were latently infected with this virus. Specifically, one group of rabbits received a transdermal patch of nicotine (21 mg/day) for 20 days and another group did not receive a patch and thus served as the control. When a tear film was collected following ocular swabbing, it was found that 16.5% (258/1,560) of the swabs taken from rabbits treated with nicotine were positive for virus, compared with 8.3% (53/639) of swabs taken from controls, strongly suggesting that a systemic exposure to nicotine significantly increases HSV-1 reactivation.[91]

6. SUMMARY

The hypothesis that nicotine alters immune responses and subsequently the health of individuals is now being investigated at many levels including molecular, cellular, and whole animal. There is now much interest in the subject of nicotine's direct and indirect effects on host immune responses, especially since it is now widely recognized that nicotine does have marked demonstrable immunomodulatory effects. For example, various laboratories have shown that nicotine may alter numerous components of the immune response system including production of inflammatory cytokines, apoptosis, and susceptibility to infection. These changes have been elucidated in experimental animals, in individuals exposed to nicotine by smoking, and by people using nicotine patches for medicinal uses including controlling addiction to smoking, increasing memory in Alzheimer patients, reducing tics in Tourette's syndrome, or modifying disease symptoms in patients with inflammatory bowel disease. A concern of the chronic exposure to nicotine is that in addition to its beneficial purposes, it may prove to have detrimental qualities. Therefore, any possible roles related to increased risk of cancer, cardiovascular disease, and respiratory infections need to be assessed. Currently it is known that nicotine influences the viability of cells through apoptosis, and the ability to produce many cytokines. These characteristics impact on how the body handles antigenic challenge, whether it is tumor exposure or infectious disease. The impact of nicotine could therefore be negative in some cases—if tumor cells were blocked in their ability to die or if immune cells were inhibited in their ability to handle an infectious agent. The impact of nicotine could be positive in other cases—if an unnecessary inflammatory response by the body needed to be controlled.

Since nicotine is now classified as an addictive substance, scientific interest into the action of nicotine actually becomes more directed. The wide availability of nicotine to individuals by cigarettes and also in therapeutic patches, liquids, and pills makes it even more urgent that further studies be performed to determine how nicotine affects biologic functions. It is apparent such further analysis of the role of nicotine in modulating host immunity and physiology will provide new information permitting a better understanding of how nicotine, as well as other addictive drugs, affect host resistance to infectious diseases.

REFERENCES

1. Bhandari, N., Sylvester, S. L. *et al.* (1996). *Source Book of Substance Abuse and Addiction.*

2. Hiemke, C., Stolp, M. *et al.* (1996). Expression of alpha subunit genes of nicotinic acetylcholine receptors in human lymphocytes. *Neurosci. Lett.* **214**(2–3):171–174.

3. Baron, J. A. (1996). Beneficial effects of nicotine and cigarette smoking: The real, the possible and the spurious. *Br. Med. Bull.* **52**(1):58–73.

4. Cinciripini, P. M., Hecht, S. S. *et al.* (1997). Tobacco addiction: Implications for treatment and cancer prevention. *J. Natl. Cancer. Inst.* **89**(24):1852–1867.

5. Domino, E. F. (1998). Tobacco smoking and nicotine neuropsychopharmacology: Some future research directions. *Neuropsychopharmacology* **18**(6):456–468.

6. Gorell, J. M., Rybicki, B. A. *et al.* (1999). Smoking and Parkinson's disease: A dose-response relationship. *Neurology* **52**(1):115–119.

7. Hughes, J. R., Goldstein, M. G. *et al.* (1999). Recent advances in the pharmacotherapy of smoking. *JAMA* **281**(1):72–76.

8. Miller, K., Hudspith, B. *et al.* (1996). Effect of cigarette smoke exposure on biomarkers of lung injury in the rat. *Inhalation Toxicol.* **8**:803–817.

9. Hoffmann, D., Djordjevic, M. V. *et al.* (1997). The changing cigarette. *Prev. Med.* **26**(4):427–434.

10. Prokopczyk, B., Cox, J. E. *et al.* (1997). Identification of tobacco-specific carcinogen in the cervical mucus of smokers and nonsmokers. *J. Natl. Cancer. Inst.* **89**(12):868–873.

11. Chowdhury, P., Doi, R., *et al.* (1993). Tissue distribution of [3H]-nicotine in rats. *Biomed. Environ. Sci.* **6**(1):59–64.

12. Henningfield, J. E., Stapleton, J. M., *et al.* (1993). Higher levels of nicotine in arterial than in venous blood after cigarette smoking. *Drug Alcohol Depend.* **33**(1):23–29.

13. Benowitz, N. (1996). Pharmacology of nicotine: Addiction and therapeutics. *Annu. Rev. Pharmacol. Toxicol.* **36:** 597–613.

14. Heller, J. and Taylor, P. (1996). *Goodman and Gilman's The Pharmological Basis of Therapeutics,* McGraw-Hill, New York.

15. Benowitz, N. L. (1997). Systemic absorption and effects of nicotine from smokeless tobacco. *Adv. Dent. Res.* **11**(3):336–341.

16. Warpman, U., Friberg, L. *et al.* (1998). Regulation of nicotinic receptor subtypes following chronic nicotinic agonist exposure in M10 and SH-SY5Y neuroblastoma cells. *J. Neurochem.* **70**(5):2028–2037.

17. Rosecrans, J. A. and Karin, L. D. (1998). Effects of nicotine on the hypothalamic-pituitary-axis (HPA) and immune function: Introduction to the Sixth Nicotine Round Table Satellite, American Society of Addiction Medicine Nicotine Dependence Meeting, November 15, 1997. *Psychoneuroendocrinology* **23**(2):95–102.

18. Nordberg, A., Fuxe, K., Holmstedt, B., and Sundwall, A. (eds) (1989). *Nicotinic Receptors in the CNS—Their Role in Synaptic Transmission,* Progress in Brain Research. Elsevier, Amsterdam.

19. Rinne, J. O., Myllykyla, T. *et al.* (1991). A postmortem study of brain nicotinic receptors in Parkinson's and Alzheimer's disease. *Brain Res.* **547**(1):167–170.

20. Yamamura, M., Amano, Y. *et al.* (1998). Calcium mobilization during nicotine-induced cell death in human glioma and glioblastoma cell lines. *Anticancer Res.* **18**(4A):2499–2502.

21. Sopori, M. L. and Kozak, W. (1998). Immunomodulatory effects of cigarette smoke. *J. Neuroimmunol.* **83**(1-2):148–156.

22. Sopori, M. L., Kozak, W. *et al.* (1998). Nicotine-induced modulation of T Cell function. Implications for inflammation and infection. *Adv. Exp. Med. Biol.* **437**:279–289.

23. Aoshiba, K., Nagai, A. *et al.* (1996). Nicotine prolongs neutrophil survival by suppressing apoptosis. *J. Lab. Clin. Med.* **127**(2):186–194.

24. Wecker, L. and Yu, Z. J. (1995). Function of nicotinic receptors in the CNS. In T.W. Stone (ed.), *CNS Neurotransmitters and Neuromodulators—acetylcholine,* CRC Press, Boca Raton, FL, pp. 105–113.

25. Whiting, P. and Lindstrom, J. (1986). Pharmacological properties of immuno-isolated neuronal nicotinic receptors. *J. Neurosci.* **6**(10):3061–3069.
26. Maslinski, W., Laskowska-Bozek, H. *et al.* (1992). Nicotinic receptors of rat lymphocytes during adjuvant polyarthritis. *J. Neurosci. Res.* **31**(2):336–340.
27. Ashakumary, L. and Vijayammal, P. L. (1997). Effect of nicotine on lipoprotein metabolism in rats. *Lipids* **32**(3):311–315.
28. Davies, B. D., Hoss, W. *et al.* (1982). Evidence for a noncholinergic nicotine receptor on human phagocytic leukocytes. *Mol. Cell. Biochem.* **44**(1):23–31.
29. Adem, A., Nordberg, A. *et al.* (1986). Extraneural cholinergic markers in Alzheimer's and Parkinson's disease. *Prog. Neuropsychopharmacol. Biol. Psychiatry* **10**(3–5):247–257.
30. Sato, K. Z., Fujii, T. *et al.* (1999). Diversity of mRNA expression for muscarinic acetylcholine receptor subtypes and neuronal nicotinic acetylcholine receptor subunits in human mononuclear leukocytes and leukemic cell lines. *Neurosci. Lett.* **266**(1):17–20.
31. Geng, Y., Savage, S. M. *et al.* (1995). Effects of nicotine on the immune response. I. Chronic exposure to nicotine impairs antigen receptor-mediated signal transduction in lymphocytes. *Toxicol. Appl. Pharmacol.* **135**(2):268–278.
32. Laskowska-Bozek, H., Bany, U. *et al.* (1996). Effect of cholinergic stimulation on free intracellular Ca2+ concentration in human lymphocytes. *Neuroimmunomodulation* **3**(4):247–253.
33. Zhang, S. and Petro, Thomas M. (1996). The effect of nicotine on murine CD4 T cell responses. *Int. J. Immunopharmacol.* **18**:467–478.
34. Carlson, N. G., Wieggel, W. A. *et al.* (1999). Inflammatory cytokines IL-1 alpha, IL-1 beta, IL-6, and TNF-alpha impart neuroprotection to an excitotoxin through distinct pathways. *J. Immunol.* **163**(7):3963–3968.
35. Hallquist, N., Hakki, A. *et al.* (2000). Differential effects of nicotine and aging on splenocyte proliferation and the production of Th1- versus Th2-type cytokines. *Proc. Soc. Exp. Biol. Med.* **224**(3):141–146.
36. Madretsma, S., Wolters, L. M. *et al.* (1996). In-vivo effect of nicotine on cytokine production by human non-adherent mononuclear cells. *Eur. J. Gastroenterol. Hepatol.* **8**(10):1017–1020.
37. Hakki, A., Hallquist, N. *et al.* (2000). Differential impact of nicotine on cellular proliferation and cytokine production by LPS-stimulated murine splenocytes. *Int. J. Immunopharmacol.* **22**(6):403–410.
38. Madretsma, G. S., Donze, G. J. *et al.* (1996). Nicotine inhibits the in vitro production of interleukin 2 and tumour necrosis factor-alpha by human mononuclear cells. *Immunopharmacology* **35**(1):47–51.
39. Hockertz, S., Emmendorffer, A. *et al.* (1994). Acute effects of smoking and high experimental exposure to environmental tobacco smoke (ETS) on the immune system. *Cell Biol. Toxico.* **10**.
40. Cohen, J. J., Duke, R. C. *et al.* (1992). Apoptosis and programmed cell death in immunity. *Annu. Rev. Immunol.* **10**:267–293.
41. Aggarwal, S. and Gupta, S. (1998). Increased apoptosis of T cell subsets in aging humans: Altered expression of Fas (CD95), Fas ligand, Bcl-2, and Bax. *J. Immunol.* **160**(4):1627–1637.
42. Kerr, J. F., Wyllie, A. H. *et al.* (1972). Apoptosis: A basic biological phenomenon with wide-ranging implications in tissue kinetics. *Br. J. Cancer* **26**(4):239–257.
43. Martin, S. J. and D. R. Green (1995). Apoptosis and cancer: The failure of controls on cell death and cell survival. *Crit. Rev. Oncol. Hematol.* **18**(2):137–153.
44. Martin, S. J. and D. R. Green (1995). Protease activation during apoptosis: Death by a thousand cuts? *Cell* **82**(3):349–352.
45. Biola, A., Andreau, K. *et al.* (2000). The glucocorticoid receptor and STAT6 physically and functionally interact in T-lymphocytes. *FEBS Lett.* **487**(2):229–233.
46. Planey, S. L. and Litwack G., (2000). Glucocorticoid-induced apoptosis in lymphocytes [In Process Citation]. *Biochem. Biophys. Res. Commun.* **279**(2):307–312.
47. Woronicz, J., Calnan, B. *et al.* (1995). Death genes in T cells. *Curr. Top. Microbiol. Immunol.* **200**:137–46.

48. Wright, S. C., Zhong, J. *et al.* (1994). Inhibition of apoptosis as a mechanism of tumor promotion. *FASEB J.* **8**(9):654–660.

49. Allen, R. T., Cluck, M. W. *et al.* (1998). Mechanisms controlling cellular suicide: Role of Bcl-2 and caspases. *Cell Mol. Life Sci.* **54**(5):427–445.

50. Lorenzen, J., Thiele, J. *et al.* (1997). The mummified Hodgkin cell: Cell death in Hodgkin's disease. *J. Pathol.* **182**(3):288–298.

51. Kitamura, Y., Shimohama, S. *et al.* (1998). Alteration of proteins regulating apoptosis, Bcl-2, Bcl-x, Bax, Bak, Bad, ICH-1 and CPP32, in Alzheimer's disease. *Brain Res.* **780**(2):260–269.

52. Williams, G. T. (1991). Programmed cell death: Apoptosis and oncogenesis. *Cell* **65**:1097–1098.

53. Volm, M. and Koomagi, R. (1999). The implications of proliferation and apoptosis for lung cancer metastasis. *Oncol. Rep.* **6**(2):373–376.

54. Kawiak, J., Hoser, G. *et al.* (1998). Apoptosis and some of its medical implications. *Folia Histochem. Cytobiol.* **36**(3):99–110.

55. Calderon, T. M., Factor, S. M. *et al.* (1994). An endothelial cell adhesion protein for mono-cytes recognized by monoclonal antibody IG9. Expression in vivo in inflamed human ves-sels and atherosclerotic human and Watanabe rabbit vessels. *Lab. Invest.* **70**(6):836–849.

56. Wick, G., Schett, G. *et al.* (1995). Is atherosclerosis an immunologically mediated disease? *Immunol. Today* **16**(1):27–33.

57. Wick, G., Perschinka, H. *et al.* (1999). Autoimmunity and atherosclerosis. *Am. Heart J.* **138**(5 Pt 2):444–449.

58. Owasoyo, J. O., Jay, M. *et al.* (1988). Impact of nicotine on myocardial neutrophil uptake. *Toxicol. Appl. Pharmacol.* **92**(1):86–94.

59. Cluette-Brown, J., Mulligan, J. *et al.* (1986). Oral nicotine induces an atherogenic lipopro-tein profile. *Proc. Soc. Exp. Biol. Med.* **182**(3):409–413.

60. Powell, J. T. (1998). Vascular damage from smoking: Disease mechanisms at the arterial wall. *Vasc. Med.* **3**(1):21–28.

61. Carty, C. S., Soloway, P. D. *et al.* (1996). Nicotine and cotinine stimulate secretion of basic fibroblast growth factor and affect expression of matrix metalloproteinases in cultured human smooth muscle cells (published erratum appears in *J. Vasc. Surg.* 1997 Apr, **25**(4):628). *J. Vasc. Surg.* **24**(6):927–934; discussion 934–935.

62. Hoffmann, D., Hecht, S. S. *et al.* (1983). Tumor promoters and cocarcinogens in tobacco carcinogenesis. *Environ. Health Perspect.* **50**:247–257.

63. Mathur, R. S., Mathur, S. P. *et al.* (2000). Up-regulation of epidermal growth factor-recep-tors (EGF-R) by nicotine in cervical cancer cell lines: This effect may be mediated by EGF [In Process Citation]. *Am. J. Reprod. Immunol.* **44**(2):114–120.

64. Chowdhury, P., Rayford, P. L. *et al.* (1998). Pathophysiological effects of nicotine on the pancreas. *Proc. Soc. Exp. Biol. Med.* **218**(3):168–173.

65. Chowdhury, P. and Rayford, P. L., (2000). Smoking and pancreatic disorders [In Process Citation]. *Eur. J. Gastroenterol. Hepatol.* **12**(8):869–877.

66. Maneckjee, R. and Minna, J. D. (1994). Opioids induce while nicotine suppresses apopto-sis in human lung cancer cells. *Cell Growth Differ.* **5**(10):1033–1040.

67. Quik, M., Chan, J. *et al.* (1994). Alpha-Bungarotoxin blocks the nicotinic receptor medi-ated increase in cell number in a neuroendocrine cell line. *Brain Res.* **655**(1–2):161–167.

68. Pratesi, G., Cervi, S. *et al.* (1996). Effect of serotonin and nicotine on the growth of a human small cell lung cancer xenograft. *Anticancer Res.* **16**(6B):3615–3619.

69. Cattaneo, M. G., D'Atri, F. *et al.* (1997). Mechanisms of mitogen-activated protein kinase activation by nicotine in small-cell lung carcinoma cells. *Biochem. J.* **328**(Pt 2):499–503.

70. Wright, S. C., Zhong, J. *et al.* (1993). Nicotine inhibition of apoptosis suggests a role in tumor promotion. *FASEB J* **7**(11):1045–1051.

71. Plummer, H. K., 3rd, Sheppard, B. J. *et al.* (2000). Interaction of tobacco-specific toxicants with nicotinic cholinergic regulation of fetal pulmonary neuroendocrine cells: Implications for pediatric lung disease [In Process Citation]. *Exp. Lung. Res.* **26**(2):121–135.

72. Rakowicz-Szulczynska, E. M., McIntosh, D. G. *et al.* (1994). Growth factor-mediated mechanisms of nicotine-dependent carcinogenesis. *Carcinogenesis* **15**(9):1839–1846.

73. Waggoner, S. E. and Wang, X. (1994). Effect of nicotine on proliferation of normal, malignant, and human papillomavirus-transformed human cervical cells. *Gynecol. Oncol.* **55**(1):91–95.

74. West, K. A., Brognard, J. *et al.* (2003). Rapid Akt activation by nicotine and a tobacco carcinogen modulates the phenotype of normal human airway epithelial cells. *J. Clin. Invest.* **111**(1):81–90.

75. Ye, Y. N., Liu, E. S. *et al.* (2003). Nicotine promoted colon cancer growth via EGFR, c-Src and 5-Lipoxygenase mediated signal pathway. *J. Pharmacol. Exp. Ther.* **20**:20.

76. Waldum, H. L., Nilsen, O. G. *et al.* (1996). Long-term effects of inhaled nicotine. *Life Sci.* **58**(16):1339–1346.

77. Meliska, C. J., Stunkard, M. E. *et al.* (1995). Immune function in cigarette smokers who quit smoking for 31 days. *J. Allergy Clin. Immunol.* **95**(4):901–910.

78. Qandil, R., Sandhu, H. S. *et al.* (1997). Tobacco smoking and periodontal diseases. *J. Can. Dent. Assoc.* **63**(3):187–192, 194–195.

79. Tang, K., Wu, H. *et al.* (1998). A crucial role for the mitogen-activated protein kinase pathway in nicotinic cholinergic signaling to secretory protein transcription in pheochromocytoma cells. *Mol. Pharmacol.* **54**(1):59–69.

80. Heusch, W. L. and Maneckjee, R. (1998). Signalling pathways involved in nicotine regulation of apoptosis of human lung cancer cells. *Carcinogenesis* **19**(4):551–556.

81. Yoshida, H., Sakagami, H. *et al.* (1998). Induction of DNA fragmentation by nicotine in human myelogenous leukemic cell lines. *Anticancer Res.* **18**(4A):2507–2511.

82. Hakki, A., Pennypacker, K. *et al.* (2001). Nicotine inhibition of apoptosis in murine immune cells. *Exp. Biol. Med. (Maywood)* **226**(10):947–953.

83. Hakki, A., Friedman, H. *et al.* (2002). Nicotine modulation of apoptosis in human coronary artery endothelial cells. *Int. Immunopharmacol.* **2**(10):1403–1409.

84. Pugh, P. C. and Margiotta, J. F., (2000). Nicotinic acetylcholine receptor agonists promote survival and reduce apoptosis of chick ciliary ganglion neurons. *Mol. Cell. Neurosci.* **15**(2):113–122.

85. Fuchs, B. A. and Pruett, S. B. (1993). Morphine induces apoptosis in murine thymocytes in vivo but not in vitro: Involvement of both opiate and glucocorticoid receptors. *J. Pharmacol. Exp. Ther.* **266**(1):417–423.

86. Nair, M. P., Schwartz, S. A. *et al.* (1997). Immunoregulatory effects of morphine on human lymphocytes. *Clin. Diagn. Lab. Immunol.* **4**(2):127–132.

87. Sharp, B. M., Roy, S. *et al.* (1998). Evidence for opioid receptors on cells involved in host defense and the immune system. *J. Neuroimmunol.* **83**(1–2):45–56.

88. Berger, F., Gage, F. H. *et al.* (1998). Nicotinic receptor-induced apoptotic cell death of hippocampal progenitor cells. *J. Neurosci.* **18**(17):6871–6881.

89. Matsunaga, K., Klein, T. W. *et al.* (2001). Involvement of nicotinic acetylcholine receptors in suppression of antimicrobial activity and cytokine responses of alveolar macrophages to Legionella pneumophila infection by nicotine. *J. Immunol.* **167**(11):6518–6524.

90. Yamaguchi, H., Friedman, H. *et al.* (2003). Involvement of nicotinic acetylcholine receptors in controlling Chlamydia pneumoniae growth in epithelial HEp-2 cells. *Infect. Immun.* **71**(6):3645–3647.

91. Myles, M. E., Alack, C. *et al.* (2003). Nicotine applied by transdermal patch induced HSV-1 reactivation and ocular shedding in latently infected rabbits. *J. Ocul. Pharmacol. Ther.* **19**(2):121–133.

7

Nicotine Receptors and Infections

YOSHIMASA YAMAMOTO

1. INTRODUCTION

Tobacco smoking is a significant risk factor in respiratory diseases including chronic obstructive lung disease and pneumonia. The bronchial alveolar lavage fluids obtained from tobacco smokers have increased number of alveolar macrophages and neutrophils.[1,2] Moreover, compared with nonsmokers, alveolar macrophages from smokers appear to be in an active state, exhibiting increased microsomal and lysosomal enzymes, elevated resting rates of glucose use, increased production of oxygen radicals and myeloperoxidase activity, and increased migration and chemotactic responsiveness.[3] However, despite this increased activity, alveolar macrophages from smokers appear to be deficient in phagocytosis and bactericidal activity.[4] Therefore, it has been conjectured that tobacco smoking may cause a disruption of normal lung immune function against respiratory infections. In fact, it is widely accepted that tobacco smoking is one of the risk factors for respiratory infections.[5–7] For instance, pneumonia caused by *Streptococcus pneumoniae*, the most common causative bacteria of community-acquired pneumonia, is accelerated by smoking.[8] Pneumonia caused by other bacteria, such as *Legionella* and *Chlamydia*, also frequently occurs in smokers.[5–7] However, little is known about the effect of tobacco components on antimicrobial activity and immune responses of alveolar macrophages.

Nicotine, a small organic alkaloid synthesized by tobacco plants, is the addictive component of tobacco. This small alkaloid acts as an agonist at the nicotinic

YOSHIMASA YAMAMOTO • Department of Basic Laboratory Sciences, Osaka University Graduate School of Medicine, Osaka, Japan.

Infectious Diseases and Substance Abuse, edited by Herman Friedman *et al.* Springer, New York, 2005.

acetylcholine receptors (nAChRs) found mainly in the central and peripheral nervous systems and on many other tissue cells throughout the body, including immune cells.[9,10] nAChRs are a family of ligand-gated, pentameric ion channels. In human, 16 different subunits (α1–7, α9–10, β1–4, δ, ϵ, γ) have been identified that form a large number of homo- and heteropentameric receptors with distinct structural and pharmacological properties.[11–13] The main function of this receptor family is to transmit signals for the neurotransmitter acetylcholine at neuromuscular junctions and in the central and peripheral nervous systems.[14,15] Although the function of nAChRs has been well investigated, the localization of this receptor in a nonnervous system suggests that nAChRs may have a nonsynaptic role. In this regard, it has been recently demonstrated that activation of nAChRs by a ligand such as nicotine resulted in alteration of immune functions, besides facilitation of cation flow.[13,16] In addition, modulation of growth of intracellular pathogens, such as *Legionella pneumophila* and *Chlamydia pneumoniae*, has also been shown following stimulation of nAChRs with ligands.[17,18] Therefore, in this chapter, a possible role of nAChRs in infection is highlighted.

2.   *LEGIONELLA PNEUMOPHILA* INFECTION AND nAChRs

L. pneumophila, an intracellular opportunistic gram-negative bacterium which infects primarily macrophages, is an etiologic cause of serious pneumonia in immunocompromised individuals, including heavy smokers.[5,6] The mechanism by which *L. pneumophila* infection of the lung is controlled is not yet clear, but it is widely accepted that the activation of macrophages to suppress intracellular bacterial growth is an essential mechanism of the resolution of the pneumonia caused by this pathogen.[19] Th1 cells are essential for the development of cell-mediated immunity and may play a pivotal role in the defense against *L. pneumophila* infection. It is known that the Th1 cytokine interferon (IFN)-γ can activate macrophages and monocytes to inhibit *L. pneumophila* growth[20] and that Th1 cells play an essential role in the development of cell-mediated immunity to pathogens.[21] Both IFN-γ and interleukin (IL)-12, which has a major role in the differentiation of the T-helper cell phenotypes, are produced by macrophages. In addition, it has been reported that the inflammatory cytokine tumor necrosis factor (TNF)-α is required for the prompt resolution of pneumonia caused by *L. pneumophila*, and a direct role for TNF-α in the activation of phagocytes has been indicated.[22] Other inflammatory cytokines, such as IL-6, are also known to control infections.[23,24] In contrast, Th2 cytokines, IL-10 in particular, may facilitate the growth of *L. pneumophila* in permissive mononuclear phagocytes, due, in part, to IL-10-mediated inhibition of TNF-α secretion and IFN-γ-mediated mononuclear phagocyte activation.[25] All these cytokines are known to be produced by macrophages in response to bacterial infections and may play a critical role in the host defense against infection.

2.1.   nAChRs of Macrophages

Although it is known that nAChRs are differentially expressed in many tissues,[26] the existence of this receptors on lung tissue and cells has not been well

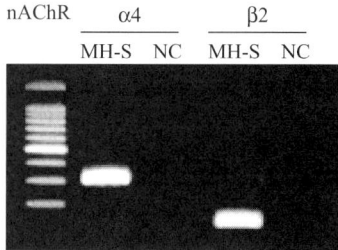

FIGURE 1. nAChR mRNA expression levels in alveolar macrophages. The expression levels of nAChR mRNAs (α4 and β2 subunits) in MH-S alveolar macrophages were analyzed by RT-PCR. NC, PCR products of MH-S cells without RT.[17]

investigated. If nAChRs exist on alveolar macrophages, the nicotine-induced immunomodulation of macrophages may be possibly mediated by nAChRs. To determine such a possibility, steady-state levels of nAChR mRNA in MH-S cells were analyzed by RT-PCR. The MH-S cells are a continuous cell line of murine alveolar macrophages, which were established after transformation of cells obtained by bronchoalveolar lavage from BALB/c mice with simian virus 40.[27] Results of characterization studies of MH-S cells indicate that this cell line may facilitate studies where homogeneous populations of alveolar macrophages are desirable, especially those involved in determining the immunologic responses of alveolar macrophages to bacterial infection. As shown in Fig. 1, mRNAs for the nAChR α4 and β2 subunits were detected in MH-S alveolar macrophages. However, it is not clear whether MH-S cells possibly may not express other nAChR subunits. Even though the results show only expression levels of certain nAChR mRNAs but not protein expression, the presence of at least α4β2-nAChRs in MH-S alveolar macrophages is highly likely.

2.2.   nAChRs and Cytokine Production

It has been shown that the treatment of human peripheral blood mononuclear cells with nicotine significantly inhibited the production of IL-2, TNF-α, and IFN-γ in response to anti-CD3 stimulation.[28] The suppression of LPS-induced murine splenocyte production of IL-6, TNF-α, and IFN-γ by concurrent nicotine treatment has also been demonstrated.[29,30] However, whether stimulation of nAChRs with nicotine as well as other specific agonists alters cytokine response of alveolar macrophages to bacterial infections, such as *L. pneumophila* infection, is not clear. In this regard, the effect of nicotine on the production of macrophage cytokines was examined (Fig. 2). The treatment of macrophages with nicotine (10 μg/ml) alone slightly induced macrophage IL-6, IL-10, IL-12, and TNF-α protein production, but this was minimal when compared with bacteria-infected macrophages. In contrast, nicotine treatment markedly downregulated the production of these cytokines, except IL-10, induced by *L. pneumophila* infection in a dose-dependent manner, even with a concentration as low as 0.1 μg/ml in the case of TNF-α. The production of IL-10 induced by *L. pneumophila* infection was not affected by nicotine, even with a concentration as high

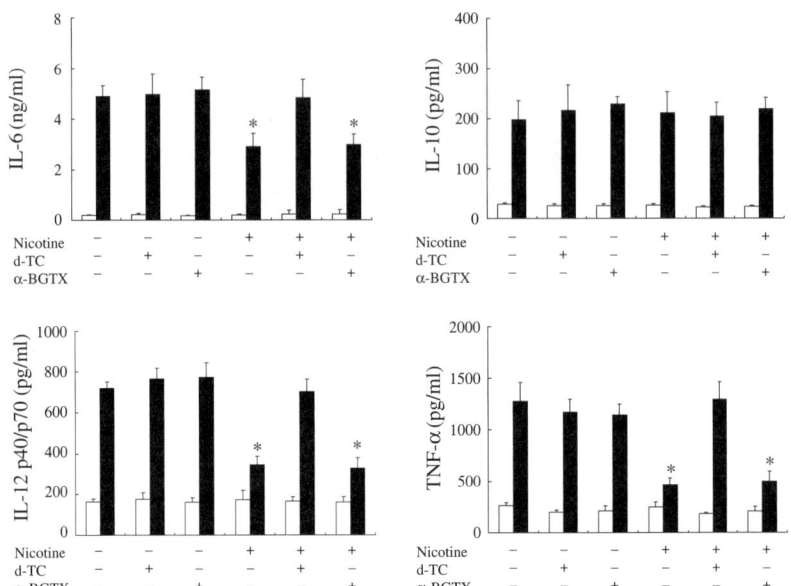

FIGURE 2. Effect of nAChR antagonists on cytokine production of nicotine-treated or untreated macrophages in response to *L. pneumophila* infection. Macrophage cultures infected with *L. pneumophila* and pretreated with or without 10 μM d-TC or 100 nM α-BGTX were incubated with or without 10 μg/ml nicotine for 24 hr. The production of cytokines in the supernatants obtained from the macrophage cultures 24 hr after bacterial infection was measured by ELISA. □, Non-*L. pneumophila* infection group; ■, *L. pneumophila* infection group. Results are expressed as means + SD for three experiments. *, $p < 0.05$, significantly different from the control group[17].

as 10 μg/ml. When macrophages infected with bacteria were pretreated with d-tubocurarine (d-TC), nonselective antagonists for nAChRs, nicotine-induced suppression of IL-6, IL-12, and TNF-α production was readily restored to the control levels without the modulation of IL-10 production. In contrast, pretreatment of macrophages with α-bungarotoxin (α-BGTX), selective antagonists for α7-nAChR, did not result in recovery of the nicotine-suppressed cytokine production. Therefore, it can be conjectured that at least α4β2-nAChRs may be a responsible receptor for the nicotine-induced selective suppression of cytokine response to *L. pneumophila* infection in MH-S alveolar macrophages. The treatment of nAChR antagonists alone did not alter the production of cytokines tested.

To determine whether stimulation of nAChRs with other agonists causes modulation of macrophage function, the effect of another nAChR agonist on immune responses of alveolar macrophages was examined. The treatment of macrophages with a nonselective nAChR agonist 1,1-dimethyl-4-phenylpiperazinium iodide (DMPP) showed the selective downregulation of cytokine production induced by *L. pneumophila* infection (Fig. 3). This selective inhibition on cytokine production by DMPP was completely blocked by d-TC treatment.

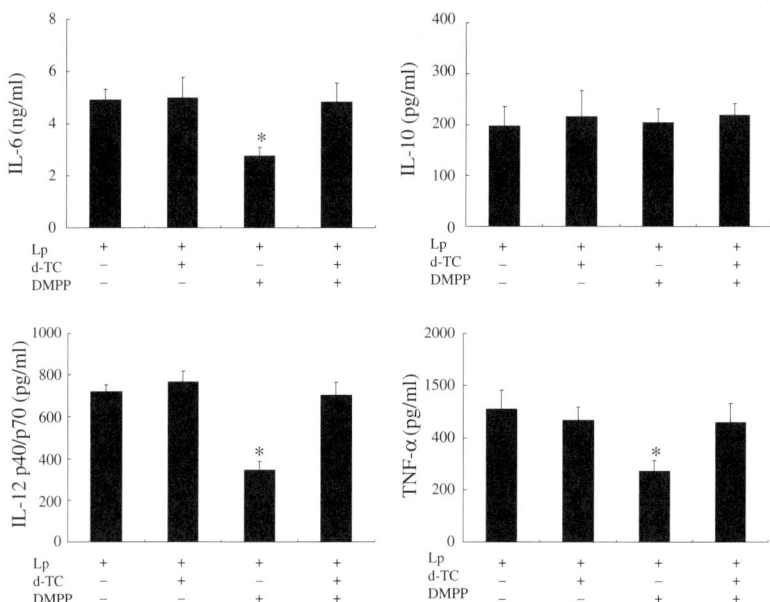

FIGURE 3. Effect of nAChR agonist DMPP on macrophage cytokine production in response to *L. pneumophila* infection. Macrophages infected with bacteria and pretreated with or without 10 μM d-TC were incubated with or without 100 μM DMPP. *, $p < 0.05$, significantly different from the *L. pneumophila* infection control group.[17]

Thus, these studies clearly indicate that the stimulation of nAChRs with nicotine as well as other agonist induces the selective downregulation of cytokine production of macrophages in response to bacterial infection. In current studies, it has also been demonstrated that nAChR α7 subunit is essential for inhibiting cytokine synthesis by the cholinergic anti-inflammatory pathway.[13] In addition, these findings have been further extended to a new avenue of research into controlling excessive inflammation.[31] That is, nicotine protects against several inflammatory diseases, such as ulcerative colitis, Parkinson's disease, and even Alzheimer's disease.

2.3. Involvement of nAChRs in Controlling *L. pneumophila* Infection

Because the growth of intracellular pathogen in macrophages is dependent on the host's macrophage activity, treatment of macrophages with nicotine, which has shown to suppress the cytokine response, may alter the growth of *L. pneumophila* in cells. The treatment of macrophages with nAChR agonist, such as nicotine as well as DMPP, after infection with bacteria induced an enhancement of the growth of *L. pneumophila* in the cells in a dose-dependent manner. Pretreatment of macrophages with antagonist d-TC completely abolished the bacteria growth enhancement of nAChR agonists (Fig. 4).

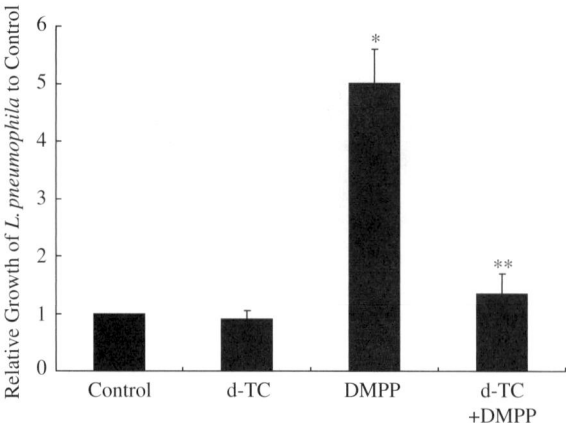

FIGURE 4. Effect of nAChR agonist on *L. pneumophila* growth in macrophages. Macrophages infected with bacteria and pretreated with or without 10 μM d-TC were incubated with or without 100 μM DMPP. *, $p < 0.05$, significantly different from the *L. pneumophila* infection control group. **, $p < 0.05$, significantly different from the DMPP-treated *L. pneumophila* infection group at the same time point. [17]

The findings of the selective inhibition of cytokine productions as shown above by stimulation of nAChRs with agonists, including nicotine, indicate how such immunomodulation may contribute to the susceptibility of cells to infections. This explanation may be supported by the fact that TNF-α is required for the prompt resolution of bacterial pneumonia and points to a direct role for TNF-α in the activation of phagocytes. [22,32] The precise mechanism of nAChR-mediated suppression of antimicrobial activity of macrophages observed is still unclear. However, our current studies have shown that epigallocatechin gallate, the major form of tea catechins, restores nicotine-suppressed TNF-α production as well as antimicrobial activity of macrophages. [33] Therefore, it seems likely that the impaired TNF-α production may be one of the major mechanisms responsible for the nAChR-mediated impairment of antimicrobial activity against *L. pneumophila* infection.

3. INVOLVEMENT OF nAChRs IN CONTROLLING *CHLAMYDIA PNEUMONIAE* GROWTH IN CELLS

C. pneumoniae is an obligate intracellular bacterium that causes a variety of respiratory illnesses, including community-acquired pneumonia, bronchitis, pharyngitis, and sinusitis. It is known that tobacco smoking accelerates pneumonia caused by *C. pneumoniae*. [8,34] In addition, the prevalence of *C pneumoniae* in clinical specimens obtained from tobacco smokers is significantly higher than that from nonsmokers. [35] Although these clinical findings indicate a possible linkage between tobacco components and acceleration of *C. pneumoniae* infection, the mechanisms of infection modulation by tobacco smoking are unclear. As observed in the study of nAChRs and *L. pneumophila* infection, a possible involvement of nAChR-mediated modulation in the susceptibility of cells to *C. pneumoniae* infection is also likely.

3.1. Nicotine and nAChR Agonists Alter *C. pneumoniae* Infection

Human epithelial HEp-2 cells are widely utilized as an *in vitro* host cell for *C .pneumoniae* infection because this pathogen preferentially infects respiratory tract epithelial cells. As shown in Fig. 5, treatment of HEp-2 cells with nicotine after infection with *C. pneumoniae* resulted in a significant increase in chlamydial inclusion numbers in cells at 72 hr after cultivation. The concentration required for significant enhancement of bacterial growth was more than 1 µg/ml, which is higher than the level in the plasma of heavy smokers (33 ± 15 ng/ml).[36] However, it has been reported that the mean nicotine yield of tobacco smoking is ~0.91 mg/cigarette.[37] Therefore, the concentration of nicotine in the respiratory tract after tobacco smoking may be higher that the level in plasma. Nevertheless, the effect of nicotine on bacterial growth was almost completely blocked by the treatment with a nonselective nAChR antagonist d-TC. The treatment of bacteria-infected HEp-2 cells with other nAChR agonists, such as acetylcholine and DMPP, also showed the significant enhancement of *C. pneumoniae* growth in cells, similar to the effect of nicotine. Furthermore, these bacterial-growth-enhancing effects of agonists were completely abolished by treatment with the antagonist d-TC, similar to the case of nicotine and d-TC treatment experiment. These results clearly show the involvement of nAChRs in the regulation of *C. pneumoniae* growth in cells.

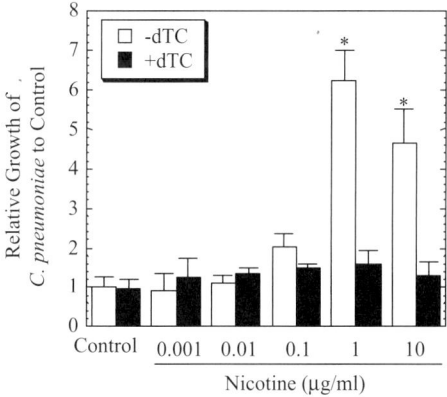

FIGURE 5. Effect of nicotine on *C. pneumoniae* growth in HEp-2 cells. Cells infected with bacteria were treated with or without the indicated concentrations of nicotine in the presence or absence of d-TC (10 µM) and then incubated for 72 hr. The number of infective progeny bacteria in cells was assessed and expressed as bacterial growth relative to that of the control group. The data shown are the mean + SD for three cultures. *, $p < 0.05$, significantly different from the control group.[18]

3.2.   nAChRs of Epithelial Cells

Even though the broad expression of nAChRs in many tissues has been recognized, it is still not clear whether and which type of nAChRs are present in epithelial HEp-2 cells. Therefore, in order to define the presence of nAChR subunit genes ($\alpha 4$, $\alpha 7$, $\beta 2$, and $\beta 4$), these were assessed by RT-PCR. The results indicate that HEp-2 cells expressed all of the subunits of nAChRs tested. Even though the study demonstrated only receptor message expression, the results of the blocking study with the antagonist and the receptor message study indicate the expression of nAChRs in HEp-2 cells. Whereas the presence of nAChRs in the cells was indicated, the type of nAChRs present in HEp-2 cells was not made clear by this study. Thus, it is obvious that nAChRs are also involved in the regulation of *C. pneumoniae* growth in cells.

4.   CONCLUSION

The widespread expression of nonneuronal acetylcholine is accompanied by the ubiquitous expression of cholinesterase and acetylcholine sensitive receptor nAChRs. Not only acetylcholine receptors but also nAChRs interact with more or less all cellular signaling pathways.[38] It is being increasingly recognized that nonneuronal acetylcholine appears to be involved in the regulation of basic cell functions, including immune functions. In this regard, the studies discussed in this chapter propose another new role for nAChRs regulation of infections by certain bacteria, such as *L. pneumophila* as well as *C. pneumoniae*, both of which are intracellular pathogens which are controlled by the host's cell function. In addition, alteration of susceptibility of cells to bacterial infection caused by stimulation of nAChRs with the exogenous ligand nicotine as well as endogenous acetylcholine indicates possible molecular mechanisms for a connection between the physiological state of a host and susceptibility to infections. Furthermore, the molecular mechanisms of tobacco smoking-induced susceptibility to infections can be also explained, at least in part, by the involvement of nAChRs. Further studies regarding involvement of nAChRs in pathophysiological aspects of infection should be performed for a better understanding of host defense, infectious diseases, and substance abuse.

REFERENCES

1. Adesina, A. M., Vallyathan, V., McQuillen, E. N., Weaver, S. O., and Craighead, J. E. (1991). Bronchiolar inflammation and fibrosis associated with smoking A morphologic cross-sectional population analysis *Am. Rev. Respir. Dis.* **143:**144–149.
2. Bosken, C. H., Hards, J., Gatter, K., and Hogg, J. C. (1992). Characterization of the inflammatory reaction in the peripheral airways of cigarette smokers using immunocytochemistry. *Am. Rev. Respir. Dis.* **145:**911–917.
3. Sopori, M. (2002). Effects of cigarette smoke on the immune system. *Nat. Rev. Immunol.* **2:**372–377.

4. Plowman, P. N. (1982). The pulmonary macrophage population of human smokers. *Ann. Occup. Hyg.* **25:**393–405.

5. Hays, J. T., Dale, L. C., Hurt, R. D., and Croghan, I. T. (1998). Trends in smoking-related diseases. Why smoking cessation is still the best medicine. *Postgrad. Med.* **104:**56–62, 65–66, 71.

6. Ruiz, M., Ewig, S., Marcos, M. A., Martinez, J. A., Arancibia, F., Mensa, J. *et al.* (1999). Etiology of community-acquired pneumonia: Impact of age, co-morbidity, and severity. *Am. J. Respir. Crit. Care. Med.* **160:**397–405.

7. Straus, W. L., Plouffe, J. F., File, T. M., Lipman, Jr., H. B., Hackman, B. H., Salstrom, S. J. *et al.* (1996). Risk factors for domestic acquisition of legionnaires disease. Ohio legionnaires Disease Group. *Arch. Intern. Med.* **156:**1685–1692.

8. Nuorti, J. P., Butler, J. C., Farley, M. M., Harrison, L. H., McGeer, A., Kolczak, M. S. *et al.* (2000). Cigarette smoking and invasive pneumococcal disease. Active Bacterial Core Surveillance Team. *N. Engl. J. Med.* **342:**681–689.

9. Drescher, D. G., Khan, K. M., Green, G. E., Morley, B. J., Beisel, K. W., Kaul, H. *et al.* (1995). Analysis of nicotinic acetylcholine receptor subunits in the cochlea of the mouse. *Comp. Biochem. Physiol. C Pharmacol. Toxicol. Endocrinol.* **112:**267–273.

10. Hiemke, C., Stolp, M., Reuss, S., Wevers, A., Reinhardt, S., Maelicke, A. *et al.* (1996). Expression of alpha subunit genes of nicotinic acetylcholine receptors in human lymphocytes. *Neurosci. Lett.* **214:**171–174.

11. Le Novere, N. and Changeux, J. P. (1995). Molecular evolution of the nicotinic acetylcholine receptor: An example of multigene family in excitable cells. *J. Mol. Evol.* **40:**155-172.

12. Leonard, S. and Bertrand, D. (2001). Neuronal nicotinic receptors: From structure to function. *Nicotine Tob. Res.* **3:**203–223.

13. Wang, H., Yu, M., Ochani, M., Amella, C. A., Tanovic, M., Susarla, S. *et al.* (2003). Nicotinic acetylcholine receptor alpha7 subunit is an essential regulator of inflammation. *Nature* **421:**384–388.

14. Marubio, L. M. and Changeux, J. (2000). Nicotinic acetylcholine receptor knockout mice as animal models for studying receptor function. *Eur. J. Pharmacol.* **393:**113–121.

15. Steinlein, O. (1998). New functions for nicotinic acetylcholine receptors? *Behav. Brain. Res.* **95:**31–35.

16. Klapproth, H., Racke, K., and Wessler, I. (1998). Acetylcholine and nicotine stimulate the release of granulocyte–macrophage colony stimulating factor from cultured human bronchial epithelial cells. *Naunyn Schmiedebergs Arch. Pharmacol.* **357:**472–475.

17. Matsunaga, K., Klein, T. W., Friedman, H., and Yamamoto, Y. (2001). Involvement of nicotinic acetylcholine receptors in suppression of antimicrobial activity and cytokine responses of alveolar macrophages to *Legionella pneumophila* infection by nicotine. *J. Immunol.* **167:**6518–6524.

18. Yamaguchi, H., Friedman, H., and Yamamoto, Y. (2003). Involvement of nicotinic acetylcholine receptors in controlling *Chlamydia pneumoniae* growth in epithelial HEp-2 cells. *Infect. Immun.* **71:**3645–3647.

19. Friedman, H., Yamamoto, Y., and Klein, T. W. (2002). *Legionella pneumophila* pathogenesis and immunity. *Semin. Pediatr. Infect. Dis.* **13:**273–279.

20. Nash, T. W., Libby, D. M., and Horwitz, M. A. (1988). IFN-gamma-activated human alveolar macrophages inhibit the intracellular multiplication of *Legionella pneumophila*. *J. Immunol.* **140:**3978–3981.

21. Hsieh, C. S., Macatonia, S. E., Tripp, C. S., Wolf, S. F., O'Garra, A., and Murphy, K. M. (1993). Development of TH1 CD4+ T cells through IL-12 produced by *Listeria*-induced macrophages. *Science* **260:**547–549.

22. Skerrett, S. J., Bagby, G. J., Schmidt, R. A., and Nelson, S. (1997). Antibody-mediated depletion of tumor necrosis factor-alpha impairs pulmonary host defenses to *Legionella pneumophila*. *J. Infect. Dis.* **176:**1019–1028.

23. Flesch, I. E. and Kaufmann, S. H. (1990). Stimulation of antibacterial macrophage activities by B-cell stimulatory factor 2 (interleukin-6). *Infect. Immun.* **58:**269–271.

24. Liu, Z., Simpson, R. J., and Cheers, C. (1992). Recombinant interleukin-6 protects mice against experimental bacterial infection. *Infect. Immun.* **60:**4402–4406.

25. Park, D. R. and Skerrett, S. J. (1996). IL-10 enhances the growth of *Legionella pneumophila* in human mononuclear phagocytes and reverses the protective effect of IFN-gamma: Differential responses of blood monocytes and alveolar macrophages. *J. Immunol.* **157:**2528–2538.

26. Cordero-Erausquin, M., Marubio, L. M., Klink, R., and Changeux, J. P. (2000). Nicotinic receptor function: New perspectives from knockout mice. *Trends Pharmacol. Sci.* **21:**211–217.

27. Mbawuike, I. N. and Herscowitz, H. B. (1989). MH-S, a murine alveolar macrophage cell line: Morphological, cytochemical, and functional characteristics. *J. Leukoc. Biol.* **46:**119–127.

28. Ouyang, Y., Virasch, N., Hao, P., Aubrey, M. T., Mukerjee, N., Bierer, B. E. *et al.* (2000). Suppression of human IL-1beta, IL-2, IFN-gamma, and TNF-alpha production by cigarette smoke extracts. *J. Allergy Clin. Immunol.* **106:**280–287.

29. Hakki, A., Hallquist, N., Friedman, H., and Pross, S. (2000). Differential impact of nicotine on cellular proliferation and cytokine production by LPS-stimulated murine splenocytes. *Int. J. Immunopharmacol.* **22:**403–410.

30. Hallquist, N., Hakki, A., Wecker, L., Friedman, H., and Pross, S. (2000). Differential effects of nicotine and aging on splenocyte proliferation and the production of Th1- versus Th2-type cytokines. *Proc. Soc. Exp. Biol. Med.* **224:**141–146.

31. Libert, C. (2003). Inflammation: A nervous connection. *Nature* **421:**328–329.

32. McHugh, S. L., Newton, C. A., Yamamoto, Y., Klein, T. W., and Friedman, H. (2000). Tumor necrosis factor induces resistance of macrophages to *Legionella pneumophila* infection. *Proc. Soc. Exp. Biol. Med.* **224:**191–196.

33. Matsunaga, K., Klein, T. W., Friedman, H., and Yamamoto, Y. (2002). In vitro therapeutic effect of epigallocatechin gallate on nicotine-induced impairment of resistance to *Legionella pneumophila* infection of established MH-S alveolar macrophages. *J. Infect. Dis.* **185:**229–236.

34. Leinonen, M., and. Saikku, P. (1999). Interaction of *Chlamydia pneumoniae* infection with other risk factors of atherosclerosis. *Am. Heart. J.* **138:**S504–S506.

35. Smieja, M., Leigh, R., Petrich, A., Chong, S., Kamada, D., Hargreave, F. E. *et al.* (2002). Smoking, season, and detection of *Chlamydia pneumoniae* DNA in clinically stable COPD patients. *BMC Infect. Dis.* **2:**12.

36. Ebert, R. V., McNabb, M. E., McCusker, K. T., and Snow, S. L. (1983). Amount of nicotine and carbon monoxide inhaled by smokers of low-tar, low-nicotine cigarettes. *JAMA* **250:**2840–2842.

37. Jarvis, M. J., Boreham, R., Primatesta, P., Feyerabend, C., and Bryant, A. (2001). Nicotine yield from machine-smoked cigarettes and nicotine intakes in smokers: Evidence from a representative population survey. *J. Natl. Cancer. Inst.* **93:**134–138.

38. Wessler, I., Kilbinger, H., Bittinger, F., and Kirkpatrick, C. J. (2001). The biological role of non-neuronal acetylcholine in plants and humans. *Jpn. J. Pharmacol.* **85:**2–10.

8

Immunomodulatory Effects of Cigarette Smoke/Nicotine

MOHAN L. SOPORI, SEDDIGHEH RAZANI-BOROUJERDI,
and SHASHI P. SINGH

1. INTRODUCTION

Cigarette smoking is a leading preventable cause of death and disability worldwide, and in the United States alone, over 400,000 deaths annually are attributed to cigarette smoking.[1] Tobacco use is linked to increased risks for atherosclerosis and heart disease; chronic obstructive pulmonary disease; respiratory track infections; periodontitis; bacterial meningitis; cancers of the lung, mouth, larynx, esophagus, and bladder; Crohn's disease; and rheumatoid arthritis, and smokers exhibit delayed recovery from injuries (reviewed in US DHHS,[2] Doll and Peto,[3] Silverstein,[4] Saag et al.,[5] Nagai et al.,[6] Sopori.[7] However, epidemiological data also suggest that smokers have a lower incidence and/or severity of some diseases, such as ulcerative colitis, sarcoidosis, endometriosis, uterine fibroids, endometrial cancer, farmers' lung, pigeon breeders' disease, Parkinson's disease, Sjögren's syndrome (reviewed in Sopori et al.,[8] Fratiglioni and Wang, [9] Manthorpe et al.,[10] Sopori[7]). Interestingly, many of these diseases are inflammatory diseases or have a significant inflammatory component.

Tobacco smoke is a complex mixture of over 4,500 chemicals, many of which have toxic and/or carcinogenic activity. In addition, many constituents of cigarette smoke, such as acrolein, benzo[a]pyrene, and hydroquinone, modulate the function of immune cells *in vitro* and/or after *in vivo* administration.[11–14] Nicotine (NT), the addictive substance in cigarettes, is a major constituent of

MOHAN L. SOPORI, SEDDIGHEH RAZANI-BOROUJERDI, and SHASHI P. SINGH •
Respiratory Immunology, Lovelace Respiratory Research Institute, Albuquerque, NM 87108.

Infectious Diseases and Substance Abuse, edited by Herman Friedman *et al.*
Springer, New York, 2005.

cigarette smoke, and cigarettes containing higher amounts of tar and NT induce immunological changes faster than cigarette smoke containing lower levels of these components (reviewed in Sopori *et al.*[8]). Thus, tar and/or NT may represent the immunosuppressive components of cigarette smoke. We, and others, have shown that NT suppresses both the adaptive and innate immune responses (reviewed in Sopori[7]; therefore, NT may contribute to the deleterious as well as the "beneficial" effects of tobacco smoke. In this chapter, we will summarize the evidence for the immunosuppressive and anti-inflammatory properties of NT.

2. NICOTINE SUPPRESSES ADAPTIVE IMMUNE RESPONSES

Chronic exposure to cigarette smoke inhibits the antibody response (reviewed in Sopori *et al.*,[8] Sopori[7]). Based on particle size, cigarette smoke is composed of two phases: the vapor phase and the particulate phase; however, chronic inhalation of the vapor phase does not affect the immune response.[15] Therefore, the particulate phase of cigarette smoke is important in cigarette smoke-induced immunosuppression. Under conditions of cigarette smoking, most of the NT is associated with the particulate phase of cigarette smoke, and animals treated chronically with NT show a significant loss of antibody-forming cell (AFC) response to sheep red blood cells (SRBC) (Table I, Exp. 1). In addition, NT suppresses T-cell mitogenesis and the migration of T cells from the G0/G1 phase into the S phase of the cell cycle.[17,18] These results are reminiscent of the changes in the AFC response in rats chronically exposed to cigarette smoke (reviewed in Sopori and Kozak[19]), suggesting that NT suppresses the immune system in a manner similar to cigarette smoke. Although NT has a very short half-life *in vivo*, the inhibition of the AFC response remains for 2–6 weeks after the removal of NT pumps (Table I, Exp. 2), indicating the development of immunological unresponsiveness in NT-treated animals.[18]

TABLE I
Chronic Nicotine Inhibits the Antibody-Forming Cell Response[a]

Treatment	Animals/group	AFC/10^6 spleen cells[a]
Exp. 1		
Control	5	762 ± 106
Nicotine (4 wk)	4	292 ± 54
Exp. 2		
Control	4	884 ± 142
Post-nicotine (2 wk)	6	234 ± 70
Post-nicotine (6 wk)	4	728 ± 79

[a]Rats were implanted subcutaneously with saline (control)- or NT-containing miniosmotic pumps; 4 days prior to sacrifice, animals were immunized with SRBC. Spleen cells were analyzed for anti-SRBC AFC responses by standard methods.[16] In Exp. 2, pumps were removed after 4 weeks of saline/NT treatment and animals sacrificed at indicated times post-NT treatment.

3. NICOTINE AFFECTS THE ANTIGEN-MEDIATED SIGNALING IN T CELLS

Stimulation of T cells through the antigen receptor (TCR) by an antigen or anti-TCR antibodies initiates a series of biochemical events that may result in T-cell proliferation, differentiation, or anergy.[20] The TCR-directed signaling in T cells can be divided into antigen recognition by the TCR complex, the cytoplasmic signal transduction cascades, and activation of the genes in the nucleus.[21] Recognition of antigens by T cells from NT-exposed animals appears normal[15]; however, following the ligation of the TCR complex with anti-$\alpha\beta$-TCR antibodies (a model for the antigen-induced T-cell activation), the major early intracellular events include the stimulation of protein tyrosine kinase activities,[22,23] leading to the activation (i.e., tyrosine phosphorylation) of phospholipase C-γ1, which cleaves phosphatidylinositol bisphosphate into inositol-1,4,5-trisphosphate (IP3) and diacylglycerol. IP3 increases the intracellular Ca^{2+} levels ($[Ca^{2+}]_i$) by releasing Ca^{2+} from the IP3-sensitive Ca^{2+} stores that in turn increases $[Ca^{2+}]_i$ by stimulating the Ca^{2+} influx.[24] The increased $[Ca^{2+}]_i$ is essential for the entry of the cell from the G0/G1 phase into the S phase of the cell cycle.

T cells from rats treated with NT for 3–4 weeks via subcutaneously implanted miniosmotic pumps exhibit reduced $[Ca^{2+}]_i$ levels in response to TCR ligation,[18] indicating that chronic NT may affect the TCR-mediated signal transduction pathway at step(s) proximal to the rise in $[Ca^{2+}]_i$. Indeed, splenic T cells from chronically NT-treated animals have constitutively stimulated protein tyrosine kinase and PLC-γ1 activities, leading to increased basal intracellular levels of IP3.[18] The constant presence of high intracellular levels of IP3 in NT-treated T cells depletes IP3-sensitive intracellular Ca^{2+} stores.[25] The ability of chronic NT exposure to promote depletion of these stores might be a major reason for the immunosuppressive effects of chronic NT exposure.

4. NICOTINE AFFECTS THE IMMUNE SYSTEM THROUGH THE CENTRAL NERVOUS SYSTEM

Increasing evidence suggests a bidirectional communication between the central nervous system and the immune system, and the two systems intimately interact during development, maturation, and aging processes (reviewed in Blalock[26]). These systems may communicate through shared signal molecules such as cytokines and neurotransmitters. Under *in vitro* conditions, lymphocytes show increased $[Ca^{2+}]_i$ in response to high concentrations of NT,[16] indicating the presence of low-affinity receptors for NT. However, chronic administration of relatively small concentrations of NT into the brain lateral ventricle causes a significant reduction in the AFC response, suggesting that some effects of NT on the immune system might be mediated through the central nervous system.[16] NT is a classical sympathoadrenal stimulant,[27] and acute NT treatment stimulates the hypothalamus–pituitary–adrenal axis, causing secretion of glucocorticoids.[28] However, our results do not support a major role for this axis in the immunosuppression caused by chronic NT exposure.[29] On the other hand, chronic NT exposure of animals pretreated with ganglionic blockers prevented the NT-induced immunosuppression (Singh *et al.*, unpublished observation). Immunological effects of low-dose sarin (nerve gas) and other cholinergic agents (i.e., physostigmine, pyridostigmine, edrophonium) are also mediated through the autonomic nervous system.[30,31] Thus, many neuroactive substances, including NT, may affect the immune system through the autonomic nervous system. To this end, T cells express adrenoceptors, which respond to norepinephrine; the latter inhibits T-cell mitogenesis.[32] These events are diagrammatically depicted in Fig. 1.

5. NICOTINE INHIBITS INFLAMMATORY AND FEVER RESPONSES

Tobacco smoking suppresses the immune system, and smokers show delayed wound repair following injuries and surgery,[4,33] prompting surgeons to advise their patients to stop smoking for a few weeks before and after surgery.[34] Interestingly, smokers have a lower incidence of some inflammatory diseases or diseases with an inflammatory component, such as ulcerative colitis, sarcoidosis, cutaneous inflammation, endometriosis, and Parkinson's disease (reviewed in Baron,[35] Eskenazi and Warner,[36] Sopori[7]). NT has been used to alleviate ulcerative colitis,[37] Parkinson's disease,[38,39] and cutaneous inflammation.[40,41] Because the inflammatory response is an important component of the innate immunity and the first line of defense against pathogens, NT treatment might encourage the growth of pathogens. Indeed, replication of influenza virus and

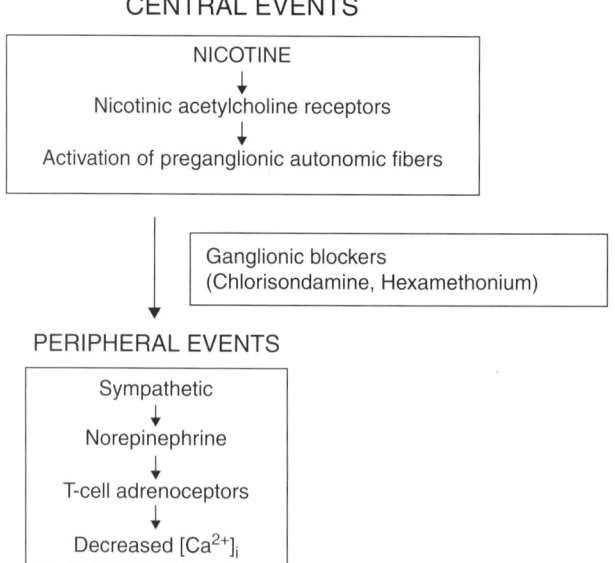

FIGURE 1. A simplified diagram showing how the effects of NT on the central nervous system might be transmitted to T lymphocytes through the autonomic nervous system.

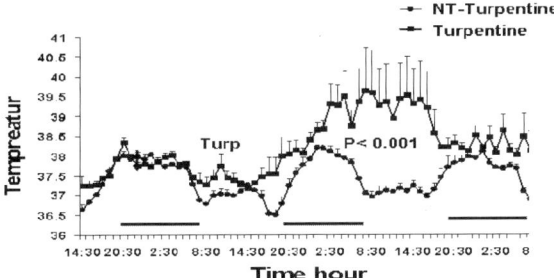

FIGURE 2. Rats were implanted with biotelemeters to record body temperature and then treated with NT. After 3 weeks, animals were injected with turpentine in the leg, and deep body temperature was recorded.

Legionella pneumophila was enhanced in the lungs of NT-treated animals and alveolar macrophage cell lines, respectively.[16,42] NT also reactivated herpes simplex virus-1 in rabbits,[43] and activation of nicotinic acetylcholine receptors on macrophages decreased their expression of proinflammatory cytokines.[44] Moreover, we have observed that a turpentine-induced increase in deep body temperature (an early event in the inflammatory response) is attenuated in NT-treated rats (Fig. 2). Thus, by suppressing inflammation in smokers, NT may retard wound repair and some inflammatory diseases, but might increase the susceptibility of smokers to pathogens.

6. SUMMARY

Studies to delineate the mechanism by which NT affects T-cell function suggest that after binding to an antigen, T cells from NT-treated animals do not normally transmit the TCR-mediated signals that would allow them to enter into the cell cycle and proliferate. Recent studies indicate that a similar defect in antigen-mediated signaling is also seen in T cells from smokers and cigarette smoke-exposed animals.[25,45] While it does not obviate the direct effects of NT on immune cells, some of the immunosuppressive effects of NT might be mediated through its effects on the brain via the autonomic nervous system. Moreover, evidence is growing that NT is an anti-inflammatory agent, which might explain the delayed wound repair process, increased susceptibility to infections, and relative resistance of smokers to some inflammatory diseases. Thus, many of the adverse as well as "beneficial" effects of smoking may result from the actions of NT on the innate and adaptive immune responses.

REFERENCES

1. CDC (Centers for Disease Control). (1999). Achievements in Public Health, 1900–1999: Tobacco Use—United States, 1900–1999. *Morb. Mortal. Wkly. Rep.* **48:** 986–993.

2. US DHHS (Department of Health and Human Services). (1982). The health conse-
    quences of smoking: Cancer. A report of the Surgeon General. Washington, DC.
3. Doll, R. and Peto, R. (1976). Mortality in relation to smoking: 20 years' observations on
    male British doctors. *BMJ* **2**:1525–1536.
4. Silverstein, P. (1992). Smoking and wound healing. *Am. J. Med.* **93**(Suppl. A1):22S–24S.
5. Saag, K. G., Cerham, J. R., Kolluri, S., Ohashi, K., Hunninghake, G. W., and Schwartz, D. A.
    (1997). Cigarette smoking and rheumatoid arthritis severity. *Ann. Rheum. Dis.* **56**:463–469.
6. Nagai, S., Hoshino, Y., Hayashi, M., and Ito, I. (2000). Smoking-related interstitial lung
    diseases. *Curr. Opin. Pulm. Med.* **6**:415–419.
7. Sopori, M. L. (2002). Effects of cigarette smoke on the immune system. *Nat. Rev. Immunol.*
    2:372–377.
8. Sopori, M. L., Goud, N. S., and Kaplan A. M. (1994). Effects of tobacco smoke on the
    immune system. In J. H. Dean *et al.* (eds), *Immunotoxicol. Immunopharmacol*, Raven Press,
    NY, pp. 413–433.
9. Fratiglioni, L. and Wang, H. X. (2000). Smoking and Parkinson's and Alzheimer's disease:
    Review of the epidemiological studies. *Behav. Brain. Res* **113**:117–120.
10. Manthorpe, R., Benoni, C., Jacobsson, L., Kirtava, Z., Larsson, A., Liedholm, R. *et al.* (2000)
    Lower frequency of focal lip sialadenitis (focus score) in smoking patients. Can tobacco
    diminish the salivary gland involvement as judged by histological examination and
    anti-SSA/Ro and anti-SSB/La antibodies in Sjogren's syndrome? *Ann. Rheum. Dis.* **59**:54–60.
11. Li, Q., Geiselhart, L., Mittler, J. N., Mudzinski, S. P., Lawrence, D. A., and Freed, B. M.
    (1996). Inhibition of human T lymphoblast proliferation by hydroquinone. *Toxicol. Appl.
    Pharmacol.* **139**:317–323.
12. Holian, A. and Li, L. (1998). Acrolein: A respiratory toxin that suppresses pulmonary host
    defense. *Rev. Environ. Health* **13**:99–108.
13. Rodriguez, J. W., Kirlen, W. G., Wirsiy, Y .G., Matheravidathu, S., Hodge, T. W., and
    Urso, P. (1999). Maternal exposure to benzo[a]pyrene alters development of T lympho-
    cytes in offspring. *Immunopharmacol. Immunotoxicol.* **21**:379–396.
14. Finkelstein, E. L., Nardini, M., and van der Vliet, A. (2001). Inhibition of neutrophil
    apoptosis by acrolein: A mechanism of tobacco-related lung disease? *Am. J. Physiol. Lung
    Cell. Mol. Physiol.* **281**:L732–739.
15. Sopori, M. L., Savage, S. M., Christner, R. F., Geng, Y. -M., and Donaldson, L. A. (1993).
    Cigarette smoke and the immune response: Mechanism of nicotine-induced immuno-
    suppression. *Adv. Biosci.* **86**:663–672.
16. Sopori, M. L., Geng, Y., Savage, S. M., Kozak, W., Soszynski, D., Kluger, M. J. *et al.* (1998).
    Effect of nicotine on the immune system: Possible regulation of immune responses by
    central and peripheral mechanisms. *Psychoneuroendocrinology* **23**:189–204.
17. Geng, Y., Savage, S. M., Johnson, L. J., Seagrave, J., and Sopori, M. L. (1995). Effects of nico-
    tine on the immune response. I. Chronic exposure to nicotine impairs antigen receptor-
    mediated signal transduction in lymphocytes. *Toxicol. Appl. Pharmacol.* **135**:268–278.
18. Geng, Y., Savage, S. M., Razani-Boroujerdi, S., and Sopori, M. L. (1996). Effects of nicotine
    on the immune response: II. Chronic nicotine treatment induces T cell anergy. *J. Immunol.*
    156:2384–2390.
19. Sopori, M. L. and Kozak, W. (1998). Mechanisms and effects of immunomodulation by
    cigarette smoke. *J. Neuroimmunol.* **81**:138–146.
20. Sloan-Lancaster, J. and Allen, P. M. (1996). Altered peptide ligand-induced partial T cell
    activation: Molecular mechanisms and role in T cell biology. *Annu. Rev. Immunol.* **14**:1–27.
22. Chan, A., Desai, D., and Weiss, A. (1994). The role of tyrosine kinases and protein tyrosine
    phosphatases in the antigen receptor signal transduction. *Ann. Rev. Immunol.* **12**:555–592.
23. Weiss, A. and Littman, D. R. (1994). Signal transduction by lymphocyte antigen receptors.
    Cell **76**:263–274.
24. Clapham, D. E. (1995). Calcium signaling. *Cell* **80**:258–268.
25. Kalra, R., Singh, S. P., Savage, S. M., Finch, G. L., and Sopori, M. L. (2000). Effects of
    cigarette smoke on the immune response: Chronic exposure to cigarette smoke impairs

antigen-mediated signaling in T cells and depletes IP3-sensitive Ca^{2+} stores. *J. Pharmacol. Exp. Ther.* **293:**166–171.

26. Blalock, J. E. (1994). The immune system, our sixth sense. *Immunologist* **2:**8–15.
27. Braubar, N. (1995). Direct effects of nicotine on the brain: Evidence for chemical addiction. *Arch. Environ. Health* **50:**263–266.
28. Seyler, L. E., Jr., Pomerleau, O. F., Fertig, J. B., Hunt, D., and Parker, K. (1986). Pituitary hormone response to cigarette smoking. *Pharmacol. Biochem. Behav.* **24:**159–162.
29. Singh, S. P., Kalra, R., Puttfarcken, P., Kozak, A., Tesfaigzi, J., and Sopori, M. L. (2000). Acute and chronic nicotine exposures modulate the immune system through different pathways. *Toxicol. Appl. Pharmacol.* **164:**65–72.
30. Kalra, R., Singh, S. P., Razani-Boroujerdi, S., Langley, R. J., Blackwell, W. B., Henderson, R. F. *et al.* (2002). Subclinical doses of the nerve gas sarin impair T cell responses through the autonomic nervous system. *Toxicol. Appl. Pharmacol.* **184:**82–87.
31. Langley, R. J., Kalra, R., Mishra, N. C., and Sopori, M. L. (2004). Central but not the peripheral action of cholinergic compounds suppresses the immune system. *J. Neuroimmunol.* **148:**140–145.
32. Kohm, A. P. and Sanders, V. M. (2000). Norepinephrine: A messenger from the brain to the immune system. *Immunol. Today* **21:**539–542.
33. Netscher, D. T. and Clamon, J. (1994). Smoking: Adverse effects on outcome for plastic surgical patients. *Plast. Surg. Nurs.* **14:**205–210.
34. Krueger, J. K. and Rohrich, R. J. (2001). Clearing the smoke: The scientific rationale for tobacco abstention with plastic surgery. *Plast. Reconstr. Surg.* **108:**1063–1073.
35. Baron, J. A. (1996). Beneficial effects of nicotine and cigarette smoking: The real, the possible and the spurious. *Br. Med. Bull.* **52:**58–73.
36. Eskenazi, B. and Warner, M. L. (1997). Epidemiology of endometriosis. *Obstet. Gynecol. Clin. North Am.* **24:**235–258.
37. Guslandi, M. (1999). Nicotine treatment for ulcerative colitis. *Br. J. Clin. Pharmacol.* **48:**481–484.
38. Ross, G. W. and Petrovitch, H. (2001). Current evidence for neuroprotective effects of nicotine and caffeine against Parkinson's disease. *Drugs Aging* **18:**797–806.
39. Mitsuoka, T., Kaseda, Y., Yamashita, H., Kohriyama, T., Kawakami, H., Nakamura, S. *et al.* (2002). Effects of nicotine chewing gum on UPDRS score and P300 in early-onset parkinsonism. *Hiroshima J. Med. Sci.* **51:**33–39.
40. Mills, C. M., Hill, S. A., and Marks, R. (1993). Altered inflammatory responses in smokers. *BMJ* **307:**91
41. Mills, C. M., Hill, S. A., and Marks, R. (1997). Transdermal nicotine suppresses cutaneous inflammation. *Arch. Dermatol.* **133:**823–825.
42. Matsunaga, K., Klein, T. W., Friedman, H., and Yamamoto, Y. (2001). Involvement of nicotinic acetylcholine receptors in suppression of antimicrobial activity and cytokine responses of alveolar macrophages to *Legionella pneumophila* infection by nicotine. *J. Immunol.* **167:**6518–6524.
43. Myles, M. E., Alack, C., Manino, P. M., Reish, E. R., Higaki, S., Maruyama, K. *et al.* (2003). Nicotine applied by transdermal patch induced HSV-1 reactivation and ocular shedding in latently infected rabbits. *J. Ocul. Pharmacol. Ther.* **19:** 121–133.
44. Wang, H., Yu, M., Ochani, M., Amella, C. A., Tanovic, M., Susarla, S. *et al.* (2003). Nicotinic acetylcholine receptor alpha7 subunit is an essential regulator of inflammation. *Nature* **421:**384–388.
45. Suzuki, N., Wakisaka, S., Takeba, Y., Mihara, S., and Sakane, T. (1999). Effects of cigarette smoking on Fas/Fas ligand expression of human lymphocytes. *Cell. Immunol.* **192:**48–53.
46. Clardy, J. (1995) The chemistry of signal transduction. *Proc. Natl. Acad. Sci. USA* **92:**56–61.

9

Regulation of Chemokine and Chemokine Receptor Expression and Function by Opioids

FILIP BEDNAR, AMBER D. STEELE, DAVID E. KAMINSKY, PENELOPE C. DAVEY, and THOMAS J. ROGERS

1. INTRODUCTION

Opioids exert a broad range of effects on immune responses either directly by altering immune cell function or indirectly by altering the expression of immune regulatory proteins such as cytokines, chemokines, and their respective receptors (reviewed in McCarthy et al.[1]). The expression of μ-, κ-, and δ-opioid receptors (MOR, KOR, and DOR, respectively) by immune cells has been definitely established on the basis of a number of parameters, including the isolation of receptor mRNA, cell binding analysis, and by flow cytometry.[2–5] The molecular basis for the immunomodulatory activities of the opioids has remained incompletely defined up to this time. However, it appears that a major mechanism of opioid-induced immunoregulation is through the control of cytokine and cytokine receptor expression and/or function. More specifically, recent reports suggest that the chemotactic cytokines (chemokines) are a significant target of the opioid-induced effects on the function of the cells of the immune system.

FILIP BEDNAR, AMBER D. STEELE, DAVID E. KAMINSKY, PENELOPE C. DAVEY, and THOMAS J. ROGERS • Center for Substance Abuse Research, Fels Institute for Cancer Research and Molecular Biology, Temple University School of Medicine, Philadelphia, PA 19140.

Infectious Diseases and Substance Abuse, edited by Herman Friedman *et al.*
Springer, New York, 2005.

The opioid and chemokine receptors are members of the G-protein-coupled seven-transmembrane receptor (GPCR) superfamily (Table I), and these receptors are involved in a number of physiological processes (Table II). Recent evidence suggests that opioid regulation of chemokine function appears to occur at two levels: chemokine and chemokine receptor *expression* and chemokine receptor *function*.

2. REGULATION OF CHEMOKINE AND CHEMOKINE RECEPTOR EXPRESSION

The high incidence of HIV infection in intravenous drug users (IVDUs) has prompted studies on the effect of opioids on HIV replication. The ability of opioids to alter HIV replication has been observed both *in vivo* and *in vitro*. Clinical studies analyzing the effect of IVDU on HIV infection have resulted in divergent results; however, these studies are confounded by polydrug abuse, differences in the doses of the opioids administered, the potential for opioid withdrawal, and patient compliance among other complications.[6,7] Similar conflicting results were observed utilizing the Simian Immunodeficiency Virus (SIV) model of HIV infection. Rhesus monkeys chronically treated with low doses of morphine were found to be less susceptible to SIV infection by SIV-smm9, a less virulent SIV

TABLE I
Summary of Selected G-Protein-Coupled Receptors and their Critical Functions

7 TM receptors	Ligands	Subgroup	Major functions
MOR	Endorphins Endomorphins	μ-opioid	Analgesia, thermoregulation, immunosuppression
KOR	Dynorphins	κ-opioid	Thermoregulation, immunosuppression, analgesia
DOR	Endorphins Enkephalins	δ-opioid	Analgesia, immunosuppression
ORL	Orphanin FQ/nociceptin	Opioid-like	Analgesia, nociception
CCR1-10	CCL1-28	CC-chemokine	Chemotaxis, HIV Coreceptors
XCR1	XCL1-2	C-chemokine	Chemotaxis
CXCR1-6	CXCL1-16	CXC-chemokine	Chemotaxis, HIV Coreceptors
CX_3CR1	CX_3CL1	CX_3C- chemokine	Chemotaxis
FPR	fMLP	Formyl-peptide	Chemotaxis, anti-bacterial immune response
FPRL1	Lipoxin A_4	Formyl-peptide	Chemotaxis

TABLE II
Major Ligands for Selected G-Protein-Coupled Receptors

7 TM receptors	Subgroups	Ligands
Opioid	MOR	Endomorphins,
	KOR	Endorphins
	DOR	Dynorphins
	OFQ/N	Enkephalins,
		endorphins
		Orphanin
		FQ/nociceptin
Chemokine	CCR1-10	CCL1-28
	XCR1	XCL1-2
	CXCR1-6	CXCL1-16
	CX$_3$CR1	CX$_3$CL1
Formyl-Peptide	FPR	FMLP
	FPRL1	Lipoxin A$_4$

strain.[7] In contrast, when the SIV-infected monkeys underwent a 2-day naloxone precipitated withdrawal from morphine, a severe, but transient immune depression was observed, followed by exacerbation of disease as recognized by higher levels of SIV$^+$CD4$^+$ T lymphocytes.[7] Chuang *et al.*[6] utilized the more virulent SIV strain mac239 along with higher doses of morphine administered chronically and found that the SIV infection was exacerbated by morphine. Therefore, apparently conflicting results may be due to differences in the virulence of the virus utilized and the amount of morphine being administered.

To determine the cause of the increase in SIV replication after chronic morphine administration, Suzuki *et al.*[8] determined that morphine elevated the expression of the HIV coreceptor, CCR5, after 24 hr and led to elevated viral binding of SIV to host cells. The ability of opioids to alter HIV coreceptor expression may provide a means for IVDUs to be more susceptible to HIV infection even if low levels of virus particles are present in a contaminated needle. Overall, the effect of opioids on SIV infection differed depending on whether dependence and tolerance to morphine was disrupted or not and the amount of morphine administered. The ability of opiates to regulate the stress response may serve to augment the latency states of SIV infection, thereby enhancing the pathogenic potential of SIV, and the stress associated with opiate withdrawal may serve to activate latent virus.[9] These findings help to elucidate factors which may confound clinical studies analyzing the effect of opiate abuse on HIV infection. The potential for polydrug use, withdrawal, and tolerance to alter individual results in epidemiological studies of IVDU and HIV-infection suggest that detailed patient histories are critical to understanding the effect of drugs of abuse on HIV infection.

Analysis of *in vitro* experimental models have provided results which suggest that opioids, such as morphine, directly alter chemokine and chemokine receptor expression. Recent work has shown that morphine, a μ-selective opioid agonist, elevated syncytia formation and reverse transcriptase activity after SIV infection of the CEMx174 human T-lymphocyte cell line.[10] Morphine was shown to enhance both mRNA and protein expression of CCR5, and this may have facilitated the elevated SIV replication.[10] The pan-opioid antagonist naloxone blocked the effects of morphine, which suggested that the morphine effect was due to activation of classical opioid receptors.[10] Methadone, a drug utilized for the treatment of opiate-dependent drug abusers, was found to elevate CCR5 expression in a naloxone-reversible manner, in monocyte-derived macrophages (MDM). Furthermore, both morphine and methadone elevated R5 HIV-1 RT expression in microglia and MDM.[11] Methadone administration to the CEMx174 cell line also elevated MOR expression and CCR5 expression at the mRNA and protein levels.[12] Therefore, methadone may further exacerbate the effects of μ-selective opioids by potentially increasing the number of cells responsive to μ-opioids or by elevating the number of responsive receptors on the cell surface.

Results from this laboratory[13] have shown that treatment with either morphine or [D-Ala^2, N-Me-Phe^4, Gly-ol^5] enkephalin (DAMGO), a μ-opioid-selective agonist, induces CXCR4 and CCR5 expression in both human CD3$^+$ lymphoblasts and CD14$^+$ monocytes. Furthermore, DAMGO-induced elevation of HIV-1 coreceptor expression is associated with an elevated replication of both X4 and R5 viral strains of HIV-1. We have suggested that the capacity of μ-opioids to increase HIV-1 coreceptor expression and replication may promote viral binding, trafficking of both HIV-1-infected cells and susceptible target cells, leading to enhanced disease progression.[13]

In contrast, MDMs treated with the κ-specific agonist, trans-3,4-dichloro-N-methyl-N[2-(1-pyrolidinyl) cyclohexyl] benzeneacetamide methanesulfonate (U50,488H), exhibited reduced HIV-1 replication, and the effect could be blocked by the κ-specific antagonist, norbinaltorphimine.[14] Similarly, U50,488H treatment reduced CD4$^+$ T-lymphocyte HIV-1 envelope fusion, and this was associated with decreased CXCR4 expression.[15] The mechanism of the κ-opioid-induced inhibition of chemokine receptor expression is uncertain. Interestingly, it appears from work carried out with murine developing T cells that the KOR does not mediate effects which are uniformly inhibitory.[16] In these studies, treatment with U50,488H was found to elevate CCR2 expression by murine thymocytes *in vitro*.

Finally, Sharp et al.[4] found that binding of the DOR by the specific δ-opioid agonist, (+)-4-((alpha R)-alpha-((2S,5R)-4-allyl-2,5-dimethyl-1-peperazinyl)-3-methoxybenzyl)-N,N-diethyl-benzamide (SNC-80), on CD4+ T lymphocytes inhibited the replication of HIV-1. The effect of SNC-80 on HIV-1 infection was blocked by a selective δ-opioid antagonist, naltrindole.[4] While it is clear that opioids seem to play a regulatory role during HIV infection, the biochemical basis for the distinct immunomodulatory effects mediated by the three opioid receptor types remains unresolved.

Opioids have also been found to alter chemokine expression. In recent studies, treatment of human peripheral blood mononuclear cells resulted in increased expression of monocyte chemoattractant protein-1 (MCP-1), regulated upon activation normal T-cell expressed and secreted (RANTES), and

IFN-α-inducible protein-10 (IP-10) expression at both the mRNA and protein levels.[17] This effect was blocked by CTAP, a μ-specific opioid antagonist, further implicating a role for the MOR.[17] The elevation of MCP-1, RANTES, and IP-10 may play a significant role in altering the chemotaxis of cells after HIV infection. Additionally, the elevation of RANTES levels, a CCR5 ligand, may inhibit viral replication by blocking viral binding to CCR5 or alter viral tropism by inhibiting R5 viral strains from infecting cells while allowing X4 viral strains to infect cells unimpeded. It should be pointed out that recent studies carried out with microglial cells show that morphine inhibits the production of RANTES by LPS and IL-1β and the effect could be blocked by naloxone.[18] The difference in cell types used and different kinetics of these experiments may explain the discrepancy between the latter results and those reported by Wetzel *et al.*[17]

Additionally, studies reported by Mahajan *et al.*[19] show that morphine treatment of either the astrocytoma cell line U87 or normal human astrocytes (NHA) inhibited IL-8 and macrophage inflammatory protein (MIP)-1β expression, while the expression of CCR3, CCR5, and CXCR2 was increased. The ability of opioids to alter the expression of chemokines and chemokine receptors by cells of the central nervous system (CNS) may dramatically affect the ability of HIV to infect microglia. Overall, opioids can alter both chemokine and chemokine receptor expression which may effect HIV-1 infection at the level of viral binding and cellular uptake (through changes in HIV coreceptor expression) and also at the level of viral target cell trafficking (through changes in chemokine expression).

2.1. Regulation of Chemokine and Opioid Receptor Function

As mentioned above, the receptors for opioids and the chemokines are both members of the GPCR superfamily. The function of these receptors can be regulated biochemically by several processes. However, the major means of GPCR control is through the process of desensitization. G-protein-coupled receptor desensitization is a ubiquitous phenomenon. The need for the regulation of signaling pathways arises when multiple signals from the extracellular environment must be integrated and coordinated to give rise to a physiological response. Broadly, this form of regulation of these receptors can be divided into homologous and heterologous desensitization. In homologous desensitization, an agonist induces the functional downregulation of its corresponding receptor in some cases resulting in its internalization and degradation or recycling. In contrast, heterologous desensitization is a regulatory mechanism which occurs between two different G-protein coupled receptors. The initiating agonist activates cellular signaling pathways through its receptor, which lead to modification and functional desensitization of an unrelated receptor. The levels of desensitization can vary ranging from functional deactivation to internalization of the second receptor. The chemoattractant family of G-protein-coupled receptors is a well-studied example of these phenomena. It is now clear that crosstalk occurs among the GPCRs utilized by a large number of diverse ligands, including the formyl peptide receptors (FPRs), complement receptors, and chemokine receptors. The precise signaling pathway(s) which mediate heterologous desensitization remain poorly defined.

A number of chemoattractants, including the formyl peptides released by bacteria, and the chemokines, induce transmembrane signaling through the G-proteins. Due to the complexity of the chemoattractant mixtures which exist *in vivo*, it is very likely that more than one set of GPCRs is active at a time during a chemotactic response. Recent literature contains several examples of chemoattractant receptor crosstalk and desensitization. Early studies focused on the modulation of FPR signaling and the consequent crosstalk with the IL-8 receptors (CXCR1 and CXCR2) or the complement component 5a receptor (C5aR) (reviewed in Ali *et al.*).[20] Similar experiments have since been undertaken with other chemokine receptors. For example, CXCR1/2 and CCR1 engage in receptor heterologous desensitization,[21] and while both CXCR1 and CXCR2 modulate CCR1 signaling, the reverse is not true. CCR1 effectively desensitizes only CXCR2, not CXCR1. CCR5 signaling desensitizes SDF-1α responses through CXCR4 in human bone marrow progenitor B cells as measured by Ca^{2+} mobilization, chemotaxis, and mitogen-activated protein kinase (MAPK) pathway activation.[22] Recent results also demonstrate heterologous desensitization of CCR5 and CXCR4 following activation of the FPR or its homologue (FPRL1).[23-25] Crosstalk of FPR with CCR5 prevents CCR5-mediated chemotaxis, Ca^{2+} mobilization, and HIV-1 entry in an agonist concentration-dependent manner. FPRL1 exhibits a broader range of signaling targets with similar desensitization phenomena occurring for both CCR5 and CXCR4. Some of these effects appear to involve protein kinase C (PKC) in the signaling pathway as inhibition with staurosporine disrupts the heterologous desensitization. With the growing number of examples of this desensitization process, the complexities underlying it are also becoming more apparent.

At this time, there appear to be multiple levels of receptor crosstalk with different pathways affecting various stages of the receptor signaling cascade. There is also an established hierarchy in the capacity of GPCRs to induce heterologous desensitization, and this appears to be inversely related to the sensitivity of a given GPCR to cross-desensitization. The two main targets of the desensitization phenomenon appear to be at the receptor/G-protein interface and the downstream signaling cascade leading to Ca^{2+} mobilization. Results show that FPR activation initiates a signaling pathway which results in the phosphorylation of both C5aR and CXCR1/2, and this is sufficient to disrupt the receptor/G-protein interaction. This leads to disruption of cellular chemotaxis, Ca^{2+} mobilization, and receptor internalization when the FPR-treated cells are exposed to either IL-8 or C5a (reviewed in Ali *et al.*).[20] FPR also induces cross-desensitization of the platelet activating factor receptor (PAFR) and the leukotriene B_4 receptor (LTB$_4$R). Reciprocal desensitization of FPR does not occur, and this may be due to the resistance of this receptor to phosphorylation.

Chemokine receptor desensitization is usually associated with target receptor phosphorylation, and this has been demonstrated recently for FPR- and FPRL1-mediated desensitization of CCR5 and CXCR4.[23-25] Whereas most responses of FPR were not desensitized by other GPCR agonists, Ca^{2+} mobilization was still sensitive to inhibition by C5a and IL-8 (reviewed in Ali *et al.*).[20]

This evidence suggests the presence of a secondary regulatory pathway involving a downstream mediator of Ca^{2+} mobilization. In comparison, this mechanism of desensitization was not observed in CCR5-mediated desensitization of FPR and FPRL1. Cells pretreated with the CCR5 agonist RANTES did not show inhibition of Ca^{2+} mobilization in response to W peptide, an FPRL1 agonist.[24] Similarly MIP-1α pretreatment did not abolish fMLF-mediated Ca^{2+} mobilization.[23] Studies of crosstalk between FPR, CCR5, and CXCR4 also demonstrated another mechanism of receptor desensitization. Activation of FPR and FPRL1 by their cognate peptide ligands led to the internalization of CCR5 (both FPR and FPRL1 mediated) and CXCR4 (only FPRL1 mediated). This process most likely represents the final step in the desensitization cascade. The presence of several different regulatory pathways for heterologous desensitization lends an additional level of complexity to the system and allows for fine-tuning of responses to distinct extracellular conditions.

The varying susceptibility of receptors to desensitization also demonstrates a signaling hierarchy within the chemoattractant receptors. FPR seems to be the most potent receptor studied thus far in terms of desensitization of other chemoattractant GPCRs. FPR activation consistently results in the downregulation of signaling through C5aR, CCR5, CXCR4, and CXCR2. Conversely, no chemotactic receptor studied in these systems has been able to completely inactivate FPR-mediated signaling. CXCR1 and C5aR are capable of desensitizing the Ca^{2+} mobilization component of FPR signaling but other signaling events induced by FPR remain intact (reviewed in Ali et al.,[20]). CCR5 and CXCR4 completely lack the ability to desensitize FPR. The general concept that seems to emerge from these signaling hierarchies is that receptors capable of heterologous desensitization are not very susceptible to the process themselves. At the opposite end, receptors that are very susceptible to heterologous desensitization often fail to induce desensitization of other GPCRs. This hierarchy might represent the priorities the cell places on the presence of each of the respective chemoattractants in the extracellular milieu. The description of the underlying signaling pathways for this hierarchy should significantly advance our understanding of the overall regulation of cellular response to the complex extracellular environment.

The biochemical mechanism(s) of heterologous desensitization appears to include both PKC-mediated target receptor phosphorylation and downstream signaling events which decrease phospholipase C (PLC) activity. The activation of the chemoattractant GPCRs leads typically to signals through the G-proteins, is usually pertussis toxin sensitive (Gi-coupled), and induces the activation of a serine threonine kinase, PKC. The activation of second messenger protein kinases inhibits the coupling of G-proteins to the receptor.

Studies were recently undertaken by Haribabu et al.[26] to determine the relationship between receptors for the chemoattractants, formyl peptides, platelet activating factor and leukotriene B_4, and the G-proteins they activated. Receptors for these chemoattractants couple to a pertussis toxin-insensitive G-protein to varying degrees. These studies showed that both pertussis toxin and wortmannin blocked ligand-induced chemotaxis. The most important finding from these studies was that all three receptors had a requirement to activate Gi in order to trigger chemotaxis. Furthermore, there appears to be a distinct G-protein usage among chemoattractant receptors, and a Gi-dependent mechanism involving Gβγ and PI3 kinase is required for chemotaxis. However, stimulation of PLC, calcium mobilization, and exocytosis can be stimulated through activation of both Gi and a pertussis toxin-insensitive G-protein.[26]

There is evidence that PKC activation is critical for the cross-desensitization between receptors that are coupled to Gi proteins. Treatment of monocytes with MCP-1 (a ligand for CCR2) or MIP-1α (ligand for CCR1 and CCR5) led to an increase in membrane Ca^{2+}-dependent and -independent PKCs (α, βI, and βII vs δ and ζ, respectively). Pretreatment with MCP-1 greatly decreased the response of monocytes to MIP-1α However, while MOR and DOR also induce cross-desensitization of CCR1, these opioid receptors activate only the calcium-independent PKCs.[27]

Receptor phosphorylation-independent cross-desensitization by activation of PLCβ also regulates chemoattractant responses. Chemoattractants fMLP, C5a, and IL-8 activate PLCβ by activating pertussis toxin-sensitive G-proteins to release Gβγ. Data has also shown that the fMLP receptor activates both PLC and inhibits cAMP production, which results in PKA activation which phosphorylates PLCβ$_3$ and blocks the ability of Gβγ to activate PLC. The PAF receptor, which couples to Gq, does not generate signals for downstream desensitization of Gi-coupled receptors.[28] However, PLCβ phosphorylation is not the only mechanism for the downstream cross-desensitization effect. Richardson *et al.*[21] showed that CXCR2 induced PLCβ phosphorylation but did not cross-desensitize fMLP or C5a receptors.

Another component in GPCR signaling that must be considered to explain the receptor phosphorylation-independent cross-desensitization is the role of regulator of G-protein signaling (RGS) proteins. There are over 20 members belonging to the RGS protein family, each containing a homologous sequence of ~125 amino acids known as the RGS box. RGS proteins function as GTPase activators for G-proteins by directly binding to Gα subunits. This results in the inhibition of G-protein signaling and a reduction in the availability of Gβγ. RGS4 expressed in a rat basophilic leukemia (RBL-2H3) cell line was shown to inhibit both homologous and heterologous phosphorylation of PAFR.[29] RGS4 was unable to inhibit FPR or CXCR1, which activate Gi, showing the selectivity of RGS4 function for receptors which couple to Gq pathways. RGS13 was found to be highly expressed in B lymphocytes and inhibited the activation of MAPK in response to CXCR4 and CXCR5 signaling in COS cells. Expression of human RGS13 or RGS3 inhibited the activation of MAPK in response to CXCR4 and CXCR5 signaling in COS cells. In addition, RGS13 and RGS1 inhibited CXCL12-induced migration of CXCR4 expressing CHO cells. RGS proteins may regulate signal length and inhibit migration of chemokines signaling through GPCRs.[30]

Since the signal transduction pathways involved in chemotaxis are not fully understood, the importance of the cytoplasmic carboxy-terminal tail in signal transduction of the MCP-1 receptor, CCR2b, has been studied. Substitution mutants of the serine or threonine residues to alanine in the carboxy-terminal tail decreased receptor internalization, but did not affect chemotaxis or signaling, as assessed by intracellular calcium mobilization or the ability to inhibit adenylyl cyclase. However, twelve amino acids (between Leu-316 and Phe-328), near the membrane proximal portion of the carboxyl terminus of CCR2b, have been found to play a role in chemotaxis, signal transduction, and agonist-dependent receptor sequestration. These results indicate that there is a dissociation of the process of chemotaxis from chemokine receptor internalization and desensitization. The relationship between signaling at the chemoattractant receptor level and cell migration is not fully understood. [31]

Crosstalk between GPCR and receptor tyrosine kinases is a complex process and the signaling molecules used for this purpose depend on both the type of receptor that is activated and the cell type studied. [32] In rat PC12 cells, the GPCR agonist bradykinnin transactivates the epidermal growth factor receptor (EGFR) and is calcium dependent. [33] In many cases, the signaling from GPCR to MAPK involves the G-protein $G\beta\gamma$ subunits which stimulate the Ras-dependent MAPK pathway. Studies have shown that the Gq-coupled thromboxane A2 receptor activates the EGFR by first activating Gi-proteins in a PKC-dependent manner. [34] These results point to a pathway where PKC activation via Gq coupling of the thromboxane receptor is followed by PKC-regulated receptor-Gi coupling, followed by EGFR activation. [34] Finally, there is some evidence that activation of MAPK may be required for homologous desensitization of the MOR. [35] These studies showed that inhibition of the MAPK pathway blocks MOR signaling and also internalization.

3. CONSEQUENCES OF HETEROLOGOUS DESENSITIZATION

It has been well established that an inflammatory response results in the generation of chemokines which are critical for summoning immune cells. Therefore, the discovery of opioid receptor expression by leukocytes, and chemokine receptor expression on neuronal cells, has generated interest in the possible interaction between these receptors subfamilies during an immune response. Numerous studies have revealed the occurrence of crosstalk between chemokine and opioid receptors on cells of the immune system, including primary human monocytes, $CD4^+$ lymphocytes, and keratinocytes, transfected Jurkat T cells, CHO cells, and RBL cells, and murine thymocytes. [15,20,36]

The presence of a hierarchy in the cross-desensitization of GPCRs has been discussed previously. This occurrence has been documented in interactions between opioid and chemokine receptors. For example, activation of CXCR4 leads to the cross-desensitization of both MOR and DOR expressed by a number of diverse cell populations. [36] On the other hand, cross-desensitization in the reverse direction is not apparent in any of these cells. These results are similar to data showing that CXCR4 desensitizes CCR7, but activation of CCR7 does not cross-desensitize CXCR4. [37] These results suggest that CXCR4 is relatively resistant to desensitization, but CXCR4 induces a strong cross-desensitization signal. In contrast, studies carried out with both primary human monocytes and CHO cells transfected with both CCR5 and MOR, suggest that the crosstalk between these GPCR is bidirectional. [38] Similarly, opioid receptors have been shown to direct desensitization toward CXCR1 and CXCR2 (receptors for IL-8), while these opioid receptors were not susceptible as targets by chemokine receptors.[39] These data not only suggest that a hierarchy for the ability to induce cross-desensitization exists, but that this appears to be inversely related to the susceptibility to heterologous desensitization.[20] These results have implications for our understanding of the function of chemokines and opioids at sites of inflammation, where both the opioids and chemokines are known to be present.

A crucial consequence of heterologous desensitization has recently been revealed by studies of the desensitization of CCR5 and CXCR4, the two major HIV coreceptors. We have found that activation of MOR leads to the desensitization of CCR5, but not CXCR4.[38] Further analysis indicates that the MOR-induced cross-desensitization is associated with reduced susceptibility to infection with CCR5-dependent HIV strains, while susceptibility to CXCR4-dependent HIV strains remains unaltered. Immunofluorescent confocal microscopy revealed that the desensitization taking place was not due to changes in cell surface expression of the chemokine receptor in primary monocytes or transfected CHO cells. Recent work shows that treatment of CD4$^+$ lymphocytes with the selective κ-agonist U50,488 led to the downregulation of membrane CXCR4 in a nor-BNI sensitive manner.[15] Moreover, U50,488 pretreatment to CD4$^+$ cells results in suppression of X4 HIV-1 (Env) glycoprotein-mediated membrane fusion.[15] These data provide evidence that the cross-desensitization induced by MOR may be mechanistically different when compared to crosstalk induced by KOR. Moreover, these data suggest that cross-desensitization may be a mechanism of inhibition of HIV uptake and infection.

Aside from the classical opioid receptors, the opioid receptor-like 1 (ORL1) receptor is widely distributed on cells of the immune system and shares greater than 40% homology with classical opioid receptors.[40] The endogenous ligand for ORL1, orphanin FQ/nociceptin (OFQ/N), has been shown to modulate a number of behavioral as well as immune responses.[41] More importantly, OFQ/N has been shown to desensitize MOR via PKC-mediated pathway.[42] Further studies are needed to more clearly define the role of ORL1 in both the immune response and heterologous desensitization.

Finally, the participation of chemokines at the neuroimmune interface remains uncertain. However, due to the phenomenon of heterologous desensitization, it is possible that ligands for CXCR4 and CCR5 can induce cross-desensitization and interrupt typical neuronal signaling necessary for pain sensation.[36] The administration of the chemokines CCL5 or CXCL12 into the periaqueductal gray (PAG) matter of the brain results in inactivation of MOR, and the loss of μ-opioid-induced analgesic activity at this site. In these studies, CCL5 or CXCL12 treatment, followed by DAMGO administration, leads to a dose-dependent reduction in DAMGO-induced analgesia.[36] These studies suggest that chemokine crosstalk with the opioid receptors may contribute to the sensation of pain at sites of inflammation, particularly in the brain.

ACKNOWLEDGMENTS. The authors wish to acknowledge support in the form of research grants from the National Institutes of Health, Institute on Drug Abuse: DA06650, DA11130, DA14230, P30DA13429, and T32DA07237.

REFERENCES

1. McCarthy, L. M., Wetzel, M. A., Sliker, J. K., Eisenstein, T. K., and Rogers, T. J. (2001). Opioids, opioid receptors, and the immune response. *Drug Alcohol Depend* **62:**111.

2. Carr, D. J., Rogers, T. J., and Weber, R. J. (1996). The relevance of opioids and opioid receptors on immunocompetence and immune homeostasis. *Proc. Soc. Exp. Biol. Med.* **213:**248.

3. Bidlack, J. M. (2000). Detection and function of opioid receptors on cells from the immune system. *Clin. Diagn. Lab. Immunol.* **7:**719.

4. Sharp, B. M., McAllen, K., Gekker, G., Shahabi, N. A., and Peterson, P. K. (2001). Immunofluorescence detection of Δ-opioid receptors (DOR) on human peripheral blood CD4+ T cells and DOR-dependent suppression of HIV-1 expression. *J. Immunol.* **167:**1097.

5. Suzuki, S., Chuang, L. F., Doi, R. H., Bidlack, J. M., and Chuang, R. Y. (2001). Kappa-opioid receptors on lymphocytes of a human lymphocytic cell line: Morphine-induced up-regulation as evidenced by competitive RT-PCR and indirect immunofluorescence. *Int. Immunopharm.* **1:**1733.

6. Chuang, L. F., Killam, Jr., K. F., and Chuang, R. Y. (1993). Increased replication of simian immunodeficiency virus in CEMx174 cells by morphine sulfate. *Biochem. Biophys. Res. Commun.* **195:**1165.

7. Donahoe, R. M., Byrd, L. D., McClure, H. M., Fultz, P., Brantley, M., Marsteller, F., *et al.* (1993). Consequences of opiate-dependency in a monkey model of AIDS. *Adv. Exp. Med. Biol.* **335:**21.

8. Suzuki, S., Chuang, A. J., Chuang, L. F., Doi, R. H., and Chuang, R. Y. (2002a). Morphine promotes simian acquired immunodeficiency syndrome virus replication in monkey peripheral mononuclear cells: Induction of CC chemokine receptor 5 expression for virus entry. *J. Infect. Dis.* **185:**1826.

9. Donahoe, R. M. and Vlahov, D. (1998). Opiates as potential cofactors in progression of HIV-1 infections to AIDS. *J. Neuroimmunol.* **83:**77.

10. Miyagi, T., Chuang, L. F., Doi, R. H., Carlos, M. P., Torres, J. V., and Chuang, R. Y. (2000). Morphine induces gene expression of CCR5 in human CEMx174 lymphocytes. *J. Biol. Chem.* **275:**31305.

11. Li, Y., Wang, X., Tian, S., Guo, C.-J., Douglas, S. D., and Ho, W.-Z. (2002). Methadone enhances human immunodeficiency virus infection of human immune cells. *J. Infect. Dis.* **185:**118.

12. Suzuki, S., Carlos, M. P., Chuang, L. F., Torres, J. V., Doi, R. H., and Chuang, R. Y. (2002b). Methadone induces CCR5 and promotes AIDS virus infection. *FEBS Lett.* **519:**173.

13. Steele, A. D., Henderson, E. E., and Rogers, T. J. (2003). μ-Opioid modulation of HIV-1 coreceptor expression and HIV-1 replication. *Virology,* **309:** 99.

14. Chao, C. C., Gekker, G., Sheng, W. S., Hu, S., and Peterson, P. K. (2001). U50,488 inhibits HIV-1 expression in acutely infected monocyte-derived macrophages. *Drug Alcohol Depend.* **62:**149.

15. Lokensgard, J. R., Gekker, G., and Peterson, P. K. (2002). Kappa-opioid receptor agonist inhibition of HIV-1 envelope glycoprotein-mediated membrane fusion and CXCR4 expression on CD4(+) lymphocytes. *Biochem. Pharmacol.* **63:**1037.

16. Zhang, L. and Rogers, T. J. (2000). Kappa-opioid regulation of thymocyte IL-7 receptor and C-C chemokine receptor 2 expression. *J. Immunol.* **164:**5088.

17. Wetzel, M. A., Steele, A. D., Eisenstein, T. K., Adler, M. W., Henderson, E. E., and Rogers, T. J. (2000). Mu-opioid induction of monocyte chemoattractant protein-1, RANTES, and IFN-gamma-inducible protein-10 expression in human peripheral blood mononuclear cells. *J. Immunol.* **165:**6519.

18. Hu, S., Chao, C. C., Hegg, C. C., Thayer, S., and Peterson, P. K. (2000). Morphine inhibits human microglial cell production of, and migration towards, RANTES. *J. Psychopharmacol.* **14:**238.

19. Mahajan, S. D., Schwartz, S. A., Shanahan, T. C., Chawda, R. P., and Nair, M. P. N. (2002). Morphine regulates gene expression of α- and β-chemokines and their receptors on astroglial cells via the opioid μ receptor. *J. Immunol.* **169:**3589.

20. Ali, H., Richardson, R. M., Haribaru, B., and Snyderman, R. (1999). Chemoattractant receptor cross-desensitization. *J. Biol. Chem.* **274**:6027.

21. Richardson, R. M., Pridgen, B. C., Haribaru, B., and Snyderman, R. (2000). Regulation of the human chemokine receptor CCR1. *J. Biol. Chem.* **275**:9201.

22. Honczarenko, H., Le, Y., Glodek, A. M., Majka, M., Campbell, J. J., Ratajczak, M. Z. *et al.* (2002). CCR5-binding chemokines modulate CXCL12 (SDF-1)-induced responses of progenitor B cells in human bone marrow through heterologous desensitization of the CXCR4 chemokine receptor. *Blood* **100**:2321.

23. Shen, W., Li, B., Wetzel, M. A., Rogers, T. J., Henderson, E. E., Su, S. B. *et al.* (2000). Downregulation of the chemokine receptor CCR5 by activation of chemotactic formyl peptide receptor in human monocytes. *Blood* **96**:2887.

24. Li, B. Q., Wetzel, M. A., Mikovits, J. A., Henderson, E. E., Rogers, T. J., Gong, W. *et al.* (2001). The synthetic peptide WKYMVm attenuates the function of the chemokine receptors CCR5 and CXCR4 through activation of formyl peptide receptor-like 1. *Blood* **97**:2941.

25. Le, Y., Wetzel, M. A., Shen, W., Gong, W., Rogers, T. J., Henderson, E. E. *et al.* (2001). Desensitization of chemokine receptor CCR5 in dendritic cells at the early stage of differentiation by activation of formyl peptide receptors. *Clin. Immunol.* **99**:365.

26. Haribabu, B., Zhelev, D. V., Pridgen, B. C., Richardson, R. M., Ali, H., and Snyderman, R. (1999). Chemoattractant receptors activate distinct pathways for chemotaxis and secretion. *J. Biol. Chem.* **274**:37087.

27. Zhang, N., Hodge, D., Rogers, T. J., and Oppenheim, J. J. (2003). Opiate receptors mediate heterologous desensitization of leukocyte chemokine receptors through Ca^{++}-independent PKCs. *J. Biol. Chem.*, **278**: 12729.

28. Hydar, A., Sozzani, S., Fisher, I., Barr, A. J., Richardson, R. M., Haribabu, B. *et al.* (1998). Differential regulation of formyl peptide and platelet-activating factor receptors. *J. Biol. Chem.* **273**:11012.

29. Richardson, R. M., Marjoram, R. J., Barr, A. J., and Snyderman, R. (2001). RGS4 inhibits platelet-activating factor receptor phosphorylation and cellular responses. *Biochemistry* **40**:3583.

30. Shi, G., Harrison, K., Wilson, G. L., Moratz, C., and Kehrl, J. H. (2002). RGS13 regulates germinal center B lymphocytes responsiveness to CXC chemokine ligand (CXCL)12 and CXCL13. *J. Immunol.* **169**:2507.

31. Arai, H., Monteclaro, F. S., Tsou, C. L., Franci, C., and Charo, I. F. (1997). Dissociation of chemotaxis from agonist-induced receptor internalization in a lymphocyte cell line transfected with CCR2B. *J. Biol. Chem.* **272**:25037.

32. Lowes, V. L., Ip, N. Y., and Wong, Y. H. (2002). Integration of signals from receptor tyrosine kinases and G protein-coupled receptors. *Neurosignals* **11**:5.

33. Zwick, E., Daub, H., Aoki, N., Yamaguchi-Aoki, Y., Tinhofer, I., Maly, K. *et al.* (1997) Critical role of calcium dependent epidermal growth factor transactivation in PC12 cell membrane depolarization and bradykinin signaling. *J. Biol. Chem.* **272**:24767.

34. Gao, Y., Tang, S., Zhou, S., and Ware, J. A. (2001). The thromboxane A2 receptor activates mitogen-activated protein kinase via protein kinase C-dependent Gi coupling and Src-dependent phosphorylation of the epidermal growth factor receptor. *J. Pharmacol. Exp. Ther.* **296**:426.

35. Polakiewiez, R. D., Schieferl, S. N., Dorner, L. F., Kansra, V., and Comb, M. J. (1998). A mitogen-activated protein kinase pathway is required for μ-opioid receptor desensitization. *J. Biol. Chem.* **273**:12402.

36. Szabo, I., Chen, X. H., Xin, L., Adler, M. W., Howard, O. M. Z., Oppenheim, J. J. *et al.* (2002). Heterologous desensitization of opioid receptors by chemokines inhibits chemotaxis and enhances the perception of pain. *Proc. Natl. Acad. Sci. USA* **99**:10276.

37. Kim, C. H., Pelus, L. M., White, J. R., Applebaum, E., Johanson, K., and Broxmeyer, H. E. (1998). CK beta-11/macrophage inflammatory protein-3 beta/EBI1-ligand chemokine is an efficacious chemoattractant for T and B cells. *J. Immunol.* **160**:2418.

38. Steele, A. D., Szabo, I., Bednar, F., and Rogers, T. J. (2002). Interactions between opioid and chemokine heterologous desensitization. *Cytokine Growth Factor Rev.* **215**:1.
39. Grimm, M. C., Ben-Baruch, A., Taub, D. D., Howard, O. M. Z., Resau, J. H., Wang, J. M. *et al.* (1998). Opiates transdeactivate chemokine receptors: δ and μ opiate receptor-mediated heterologous desensitization. *J. Exp. Med.* **188**:317.
40. Harrison, L. M. and Grandy, D. K. (2000). Opiate modulating properties of nociceptin/orphanin FQ. *Peptides* **21**:151.
41. Peluso, J., Gaveriaux-Ruff, C., Matthes, H. W. D., Filliol, D., and Keiffer, B. (1998). Distribution of nociceptin/orphanin FQ receptor transcript in human central nervous system and immune cells. *J. Neuroimmunol.* **81**:184.
42. Mandyam, C. D., Thakker, D. R., Christensen, J. L., and Standifer, K. M. (2002). Orphanin FQ/nociceptin-mediated desensitization of opioid receptor-like 1 receptor and μ opioid receptor involves protein kinase C: A molecular mechanism for heterologous cross-talk. *J. Pharmacol. Exp. Ther.* **302**:502.

10

Morphine, Th1/Th2 Differentiation, and Susceptibility to Infection

SABITA ROY, JING-HUA WANG, and
RODERICK A. BARKE

1. INTRODUCTION

Opioid abuse is a major public health problem and a controversial social issue imparting considerable economic and personal costs to societies both in the United States and internationally. It is now widely recognized that chronic opioid abuse markedly alters immune responses in humans and in experimental animal models, and thus may place the abuser at higher risk for contracting certain diseases. The idea that opioids can affect immune functions is not entirely new. As early as 1898, the effect of opium on leukocyte phagocytosis was described in a guinea pig model.[1] More recently, evidence supporting the role of opioids in suppressing a variety of immunological end points in opioid addicts has been reported by several investigators.[2-5] In animal models as well, morphine—the most commonly used opioid clinically—has also been shown to alter a number of immune parameters. The effect of morphine on immune cells is summarized in Table I.

SABITA ROY, JING-HUA WANG, and RODERICK A. BARKE • Departments of Pharmacology and Surgery, University of Minnesota, Minneapolis, MN 55455.

Infectious Diseases and Substance Abuse, edited by Herman Friedman *et al.*
Springer, New York, 2005.

TABLE I
Effect of Morphine on Immune Cells

Cell type	Parameter studied	Opioid treatment	Effect	References
Macrophage	1. Chemotactic activity	*In vivo*	Increased	[6]
	2. Opsonization	*In vitro*	Decreased	[6]
	3. TGF-β gene expression	*In vitro*	No change	[6]
	4. NO production	*In vitro*	Decreased	[6]
		In vivo	Decreased	[6]
	5. IL-1, TNF-α production	*In vitro*	Decreased	[6]
		Chronic, *in vivo*	Increased	[7,8]
	6. IL-12 production	Acute, *in vivo*	Decreased	[9]
		Chronic, *in vivo*	Increased	[7,8]
	6. IL-10 production	Chronic, *in vivo*	Decreased	[8]
		Acute, *in vivo*	Decreased	[8]
	7. β-chemokines	*In vitro*	Decreased	[10]
	8. CCR-5 receptor	*In vitro*	Increased	[10,11]
	9. apoptosis	*In vitro*	Increased	[10,11]
	10. Phagocytosis	*In vivo*	Decreased	[6]
	11. Chemokinesis	*In vivo*	Decreased	[6]
	12. IgG 1 uptake	*In vivo*	Increased	[6]
	13. [³H] Arachidonic acid uptake	*In vivo*	Increased	[6]
	14. [³H] Morphine binding			[6]
	15. Receptor cloning			[6]
NK cell	1. NK cell activity	*In vivo*	Decreased	[6]
	2. Metastatic enhancement	*In vivo*	Increased	[6]
T cell	1. Cell surface marker expression	*In vivo*	Decreased	[6]
	2. T-helper function	*In vitro*	Decreased	[6]
	3. CD4+/CD8+ population	*In vivo*	Decreased	[6]
	4. Apoptosis	*In vivo*	Increased	[6]
		In vitro	Increased	[12–14]
	5. Cell viability	*In vivo*	Decreased	[6]
	6. % of Thy1+ cells	*In vivo*	Decreased	[6]
	7. Calcium induction	*In vitro*	Decreased	[6]
	8. Cytotoxic T-lymphocyte activity	*In vitro*	Decreased	[6]
	9. Proliferative response	*In vivo*	Decreased	[6]
		In vitro	Decreased	[6]
	10. IL-2 synthesis	*In vivo*	Decreased	[6]
		In vitro	Decreased	[6]
	11. IFN-γ production	*In vivo*	Decreased	[7]
		In vitro	Decreased	[15,16]
	12. Expression of morphine binding site			[6]
	13. Expression of classical receptors			[6]
	14. Expression of opioid peptides			[6]
	15. Expression of orphan receptor			[6]

(continued)

TABLE I (continued)

Cell type	Parameter studied	Opioid treatment	Effect	References
	16. Transcripton factor NFAT	*In vivo*	Increased	[15]
	17. Expression of mu-opioid receptors	*In vitro*	Increased	[17,18]
B Cell	1. Antibody production (PFC)	*In vivo*	Decreased	[6]
	2. Polyclonal IgG production	*In vivo*	Increased	[6]
	3. Mitogenic response	*In vivo*	Decreased	[6]
	4. Induction of micronuclei	*In vivo*	Increased	[6]
	5. β-endorphin binding			[6]
	6. Opioid peptide expression			[6]
	7. Orphan receptor expression			[6]

2. ROLE OF CYTOKINES IN INFECTION

Cytokines have been recognized as key factors in determining host resistance to infectious pathogens. In particular, Th1–Th2 cytokine imbalance in hosts is associated with increased infection by intracellular microbes. The outcome of microbial infection in an organism is a dynamic process that depends on factors derived from both the microorganism and the host. In chronic human infections, specific immune response to pathogens may be of vital importance to host defense. On the other hand, an inappropriate immune response may result not only in lack of protection, but may also contribute to disease severity. Bacteria represent a heterogeneous family of pathogens that in a very simplified scheme can be grouped as either, toxin-producing bacteria, extracellular bacteria, or intracellular bacteria. Depending on the type of bacterial infection, the host mounts a specific type of immune surveillance. In the case of toxin-producing bacteria, bacterial toxin neutralization is the course of action rather than elimination of the pathogen. Bacterial toxins are neutralized by specific antibodies generated in the host through a humoral immune response. Extracellular gram-negative bacteria and gram-positive cocci and many enterobacteria cause an acute type of disease soon after host invasion and induce colonization and invasion. Specific antibodies directed against the pathogens result in the bacteria being opsonized, phagocytozed, and rapidly killed. Intracellular bacteria (e.g., *Listeria monocytogenes*, Mycobacteria, Salmonellae) are capable of surviving within mononuclear phagocytes or other host cells, which makes these pathogens insensitive to antibody-mediated elimination and enables T lymphocytes to be central to protection through activation of antibacterial capacities in the infected macrophages. In most cases, the immune response against intracellular bacteria is of the Th1 type, and depletion of CD4+ T cells or neutralization of IFNγ by monoclonal antibodies exacerbates many experimental infections induced by intracellular bacteria. An optimum host response against such diverse microbial strategies demands highly specialized reactions which are primarily controlled by CD4+ Th subsets.

2.1. Role of Th1/Th2 Differentiation in Host Defense

Cell-mediated immunity is the effector function of T lymphocytes and serves as the defense mechanism against microbes that survive within phagocytes or infect nonphagocytic cells.[19] There are two main forms of cell-mediated immune response. In the first type, delayed hypersensitivity, CD4+ Th1 cells recognize microbes that have been phagocytized by phagocytes and activate phagocyte killing mechanisms.[19] Activated macrophages kill phagocytized and extracellular microbes by generating reactive oxygen intermediates, nitric oxide, and lysosomal enzymes. In the second type of cell-mediated immunity, CTLs kill nucleated cells that contain foreign antigens. In this chapter, we will focus on the role of morphine in cellular immunity as expressed by CD4+ T-cell differentiation and not address opioid regulation of CTL-mediated actions. Appropriate induction of a Th1 differentiation is necessary for an effective response to intracellular pathogens and involves macrophage activation and production of complement fixing and opsonizing antibodies.[19] Observationally, Th1 cells organize responses to pathogens that have overcome epithelial borders and invade internal tissues.[20] Clinically, an example of the implication of impaired Th1 differentiation is infection with *Mycobacterium tuberculosis*, which is common in opioid addicted drug users.[21] Effective host defense to *M. tuberculosis* requires IFNγ synthesis, macrophage activation, and Th1 differentiation. Late progressive murine tuberculosis is accompanied by a clear switch to a Th2 dominated pattern of cytokine production. Stress has been shown to contribute to these effects, but the role of endogenous opioids in this function has not been investigated extensively.

3. CONTROL OF CD4+ T-CELL DIFFERENTIATION

The Th1 and Th2 subsets of CD4+ T-effector cells produce characteristic cytokines. Typical Th1 cytokines include IFNγ, TNFβ, and IL-2. IFNγ is the signature cytokine of Th1 cells. Th2 cytokines include IL-4, IL-5, IL-6, IL-9, IL-10, and IL-13. IL-4 is the defining cytokine for Th2 cells. Factors that control CD4+ T-cell differentiation include antigen dose, antigen-presenting cells and the cytokines they produce, host genetic background, activity of costimulatory molecules (B7-1/CD28) and hormones present in the local region (i.e., glucocorticoid).[19,22] It is generally agreed, however, that one of the most important mechanisms in CD4+ T-cell differentiation is cytokine environment. IL-12 and IFNγ are the principal cytokines driving naïve CD4+ T-cells (Th0) to Th1 differentiation. IL-4 is produced by Th2 cells and is the key factor that drives uncommitted, bipotential Th0 cells into the Th2 pathway. These signals reinforce or inhibit the expression of the canonical master regulators T-bet and GATA3.

3.1. Role of the Transcription Factor T-Bet in IFNγ Transcription and Th1 Differentiation

It is useful to consider CD4+ T-cell differentiation as a temporal process. Naïve T-helper cells represent the earliest differentiation process and either Th1 or Th2 cells represent the latter phase of the differentiation process. Figure 1 shows a simplified model of regulation of CD4+ T-cell differentiation to Th1 cells and the role of the transcription factor T-bet and GATA3. In the naïve CD4+ T cell, IL-12 induces STAT4, which is thought to (1) induce early T-bet expression, (2) prolong IFNγ synthesis, (3) induce survival and cell division, and (4) antagonize STAT6 function.[23,24] Although a number of transcription factors play an important role in the regulation of IFNγ (NFAT, NFκB families, IRF-1, c-Jun/ATF2, c-Rel, STAT4 dependent factors), studies have found no evidence for their mediation in Th1-restricted expression of IFNγ.[25-30] An important advance in our understanding of the control of Th1 lineage commitment and IFNγ expression is the identification of the Th1-restricted transcription factor protein, T-bet.[31] T-bet, a T-box family transcription factor, is now recognized as a key switch in the control of Th1 lineage commitment. Data supporting the role of T-bet in Th1 differentiation include: (1) T-bet specifies Th1 effector fate by targeting chromatin remodeling to individual IFNγ alleles, (2) T-bet silences IL-4 expression independent of IFNγ, (3) T-bet induces IL-12Rβ2 expression,

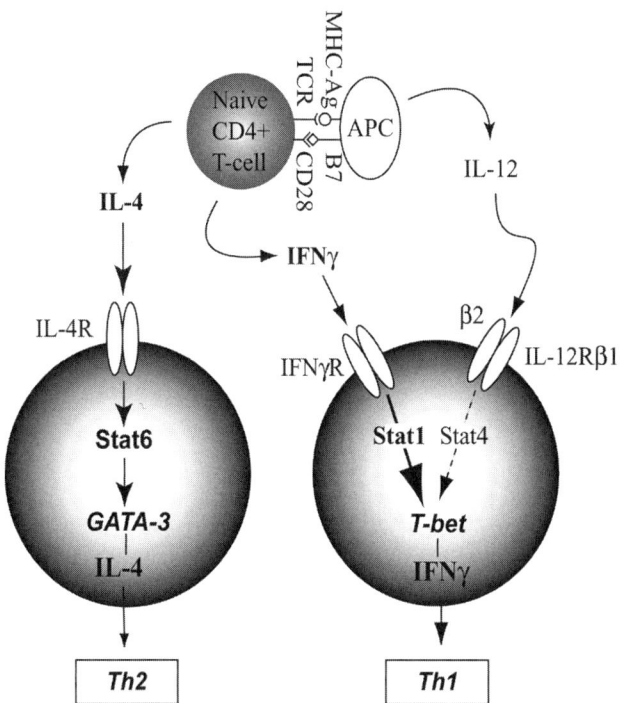

FIGURE 1. Regulation of CD4+ T-cell differentiation.

and (4) T-bet autoregulates itself similar to GATA3. Our studies show that chronic morphine treatment inhibits T-bet expression in a dose-dependent manner (Fig. 2). Our data suggests that inhibition of T-bet expression may be a possible mechanism by which chronic morphine treatment polarizes T-helper cells to Th2 differentiated effector cells.

3.2. The Role of the Transcription Factor GATA3 in Th2 Differentiation

Reprogramming of the expression of multiple cytokine genes must occur during differentiation of naïve T cells into either Th1 or Th2 effector cells. Two *Th2 tissue-specific transcription factors* have been identified: (1) the zinc-finger transcription factor GATA3 and (2) the c-Maf proto-oncogene.[31–36] GATA3 is a member of the GATA family of zinc finger proteins. Naïve Th cells (Th0 cells) express negligible levels of GATA3. Under Th2 bias conditions, GATA3 is rapidly induced. Several *in vitro* studies have demonstrated that GATA3 is sufficient in directing developing and polarized Th cells to produce Th2 cytokines.[37–39] Although other nontissue specific transcription factors are important for differentiation, they have a limited contribution to the decision process in Th1 vs Th2 differentiation.

It has become clear that GATA3 is a key switch in the control of Th2 lineage commitment. Data supporting this include the following: (1) The presence of GATA3 sites throughout the type 2 cytokine cluster, and the ability of GATA3 to (2) induce chromatin remodeling at the locus, (3) autoregulate its own expression, and (4) suppress IFNγ expression independent of IL-4/STAT6 signaling by

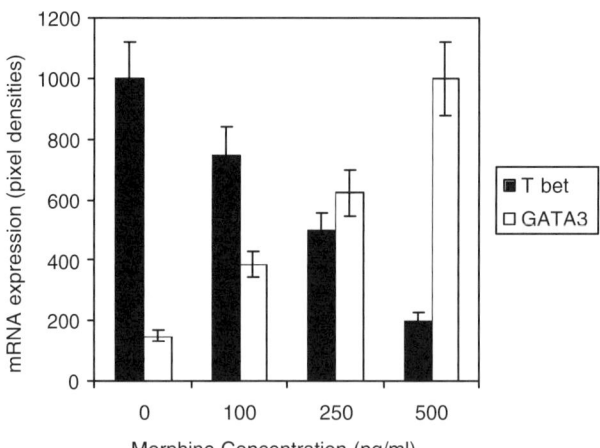

FIGURE 2. Effect of chronic morphine treatment on GATA3 and T-bet mRNA expression. Naïve splenocytes were pretreated with either vehicle or morphine (250 ng/ml) and stimulated with plate bound anti-CD3/CD28 for 72 hr. At the end of the incubation period, CD4 T cells were prepared and re-stimulated with plate bound anti-CD3/CD28 for 24 hr. Total RNA was prepared and T-bet and GATA3 mRNA levels were determined using RT-PCR.

a cell intrinsic mechanism. However, the mechanisms by which GATA3 regulates Th2 differentiation are still not very clear.[3,40,41]

Interestingly, it has been recently shown that IL-4 treatment increases μ-opioid receptor promoter activation through a STAT6 binding site.[42] In Th2 differentiated cells, this could be a possible "autocrine" mechanism by which the μ-opioid receptor regulates its own expression. Our studies show that chronic morphine treatment upregulates GATA3 expression in a dose-dependent manner (Fig. 2). From these data, we speculate that chronic morphine treatment differentially modulates the transcriptional "switches," GATA3 and T-bet, which then polarize CD4+ T cells to Th2 effector cells.

4. OPIOIDS AND Th1/Th2 DIFFERENTIATION

4.1. β-Endorphin

The *in vivo* administration of β endorphin to rodents has been shown to decrease cellular immune functions such as mitogen-induced lymphocyte proliferation and natural killer (NK) cell activity.[22] Continuous infusion of β-endorphin prolong skin allograft survival time in mice, and naloxone (opioid antagonist) administration at the time of transplantation accelerates graft rejection coinciding with an increase in splenocyte, IL-2, and IFNγ production. Blockade of β-endorphin activity either with the administration of the opioid antagonist or with β-endorphin neutralizing immunoglobulins induces an increase in both NK activity and lymphocyte proliferation.[43,44] Interestingly, it has been shown[44] that, in mice immunized with keyhole limpet hemocyanin (KLH), the vehicle treatment *in vivo* resulted in inhibition of IL-2 and IFNγ synthesis and increased IL-4 synthesis. In contrast, treatment with the opioid receptor antagonist naloxone resulted in increased IL-2 and IFNγ synthesis and inhibition of IL-4 synthesis consistent with a shift from Th2 to Th1 differentiation. These investigators hypothesized that endogenous opioids (β-endorphin?) may polarize CD4+ T-cell differentiation toward Th2 in this model. In line with this observation, Moynihan *et al.*[45] have shown that stress-odor-induced increase in splenic IL-4 production is blocked by the opioid antagonist naltrexone suggesting the role of endogenous opioids in this function. Similarly, chronic restraint stress-induced lymphocyte apoptosis and lymphocyte number are also blocked by naltrexone.[45] Consistent with these observations are studies showing an increase in endogenous opioid levels following a stress/surgical procedure.[46–48] The inhibitory effect of β-endorphin on immune function is independent of steroids. The increase in peripheral blood mononuclear cells (PBMC) and splenocyte proliferation caused by naloxone occurs despite the presence of high concentrations of corticosteroids[47] and adrenalectomy does not abolish β-endorphin's inhibitory effects.[46] The role of β-endorphin in stress-induced Th2 differentiation is further supported by the observation that the opioid peptides α- and β-endorphin are present in *in vitro* cultures of purified Con A stimulated CD4+ T cells.[49] These investigators further showed that addition of β-endorphin to CD4+ T cells resulted in a 3-fold increase in IL-4 production, and that this effect may be mediated through a naloxone sensitive site.[50,51]

4.2. Chronic Morphine and Th2 Differentiation

We have recently shown that morphine operates through a naloxone-sensitive opioid receptor mechanism to bias naïve murine CD4+ T-cell differentiation to a Th2 pathway.[52] Chronic morphine treatment temporally inhibits Th1 cytokines IL-2 and IFNγ, and increases Th2 cytokine IL-4 and IL-5 both at the transcription and protein synthesis level.[52] These effects were abolished in a μ-opioid receptor knockout mice implicating a distinct role for the μ-opioid receptor in this function. In addition, we showed that a clinical dose of morphine (4 mg/kg) superimposed upon a lipopolysaccharide (LPS)-induced infection model (an animal model of sepsis) resulted in a significant increase in mortality at 48 hr. In the absence of the drug, most septic animals died after 96 hr. Phenotypic responses, such as decreased thymic cellularity, compromised mitogenic response, and inhibition of IL-2 synthesis, that are evident at 48–72 hr after LPS injection appeared as early as 24 hr in animals that received morphine in addition to LPS. In addition, our results show that, in T cells, there is a shift from Th1-type cytokine elaboration to a Th2-type cytokine elaboration in animals that receive both LPS and morphine.[53] More recently, we have shown that chronic morphine treatment resulted in two phases of mortality when infected with a sublethal dose of *Streptococcus pneumonia* (unpublished data). The first phase occurred around day 2 and the second phase of mortality occurred around day 5. Interestingly, the second phase of mortality correlated with a shift from Th1- to Th2-producing cytokines.

5.   MECHANISMS INVOLVED IN MORPHINE-INDUCED DIFFERENTIATION OF CD4+ T CELLS TO Th2 EFFECTOR CELLS

5.1. Cyclic AMP and Th2 Differentiation

cAMP is an important second messenger with immunomodulatory properties. In effector T cells, an increase in the level of intracellular cAMP inhibits cytokine production in Th1 cells but stimulates cytokine production in Th2 cells.[54] It has been shown that cAMP-induced effects in Th2 cells may occur independently of the protein kinase A (PKA) pathway, which is the major mediator of cAMP-induced signaling events in most cell types.[54] cAMP causes increased phosphorylation of the transcription factor GATA3. Interestingly cAMP has been shown to inhibit JNK and ERK activity in Th0-like cell lines (Jurkat and EL4); however, in fully differentiated Th2 cells (D10 cells), cAMP stimulates p38 mitogen-activated protein kinase (MAPK). This effect appears to be Th2 selective since cAMP has little effect on p38 phosphorylation in Th1 cells.

5.2. Morphine and Adenyl Cyclase Superactivation

Morphine has been demonstrated to modulate cell function, in part, through a cAMP-mediated mechanism. Morphine temporally regulates adenyl cyclase. Based on studies in neuronal cell types, acute stimulation of opioid receptors activates the $G\alpha_{i/o}$ GTP-binding proteins, resulting in acute inhibition of adenyl cyclase and reduction in camp production.[55–58] Chronic opioid exposure in neuronal cell types gradually leads to molecular and cellular adaptations that result in upregulation of the cAMP pathway (adenyl cyclase superactivation or overshoot).[55–57,59,60] A study of the μ-opioid receptor specific agonist DAMGO ([d-Ala^2, N-methyl-Phe^4, gly-ol^5] enkephalin)[61,62] found that while acute DAMGO treatment (10 min) inhibited adenyl cyclase activity, chronic DAMGO treatment resulted in adenyl cyclase superactivation. Interestingly, adenyl cyclase superactivation depended on constant stimulation of the receptor. Both acute inhibition and superactivation of adenyl cyclase by morphine was shown to be antagonized by pertussis toxin pretreatment. We show that chronic treatment of splenocytes with morphine results in an increase in forskolin-induced adenyl cyclase activity (Table II). Similar to the observation in neuronal cells, morphine-induced increase in adenyl cyclase activity was abolished when cells were pretreated with pertussis toxin. Furthermore, we demonstrated that morphine-induced increase in IL-4 and decrease in IFNγ can be antagonized by pretreatment with pertussis toxin (Table II).

Based on these studies, we speculated that chronic morphine treatment polarizes naïve CD4+ T-cell differentiation toward Th2 through an adenyl cyclase mediated mechanism. Morphine acting through cAMP may enhance this process by either (1) altering naïve T-helper cell reprogramming by modulating the transcription factor GATA3, or (2) inhibiting T-bet and thereby shifting the balance of Th1 vs Th2 cytokine production. This relationship between the second messenger cAMP and Th2 differentiation has been suggested by Novak et al., [63] who demonstrated that representative Th2 cell lines maintain significantly higher levels of cAMP per cell than Th1 cell lines. Lee et al.,[64] using a GATA3 overexpression model, demonstrated that ectopic expression of GATA3 induces Th2-specific cytokine expression not only in developing Th1 cells but also in otherwise irreversibly committed Th1 cells. Moreover, cAMP markedly augmented Th2 cytokine production in GATA3-expressing Th1 cells.[63]

TABLE II
Effect of Pertussis Toxin on Morphine-Induced Adenylate
Cyclase Activity, IL-4 and IFNγ Expression

Treatment groups	Adenyl cyclase activity (%)	IL-4 (%)	IFNγ (%)
Vehicle	100	100	100
Morphine (250 ng/ml)	275	225	37
Morphine (250 ng/ml) + PTX	115	120	92
Vehicle + PTX	98	102	96

These authors suggested that cAMP elevating agents may induce a switch of lymphokine production toward Th2 phenotype.

5.3. CREB and Th2 Differentiation

Chronic morphine treatment increases levels of adenyl cyclase and cAMP-dependent protein kinase activity in the locus coeruleus (LC).[59] The transcription factor CREB has been implicated in these effects. Reduction of CREB immunoreactivity in the LC, achieved by infusion of CREB antisense oligonucleotide, completely blocked the morphine-induced upregulation of adenyl cyclase, though not of PKA.[59] The oligonucleotide effect was sequence specific. Consistent with this result, we showed that chronic morphine treatment of splenocytes induced an increase in the binding of the transcription factor CREB to its consensus DNA oligonucleotide in an electromobility shift assay (Fig. 3). Future studies will focus on the mechanisms by which morphine induces an increase in CREB regulation of adenyl cyclase and PKA in immune cells, and the role of these factors in modulation of T-bet and GATA3 expression in CD4+ T cells.

6. MORPHINE REGULATION OF MAPKs

Several investigations have implicated the MAPK/ERK pathway in response to morphine treatment in brain, lymphocyte cells, models of angiogenesis, and in COS-7 cells. Chronic, systemic administration of morphine results in a sustained increase in ERK phosphorylation state and activity in the ventral tegmental area of the brain.[65,66] Chronic morphine exposure also increased

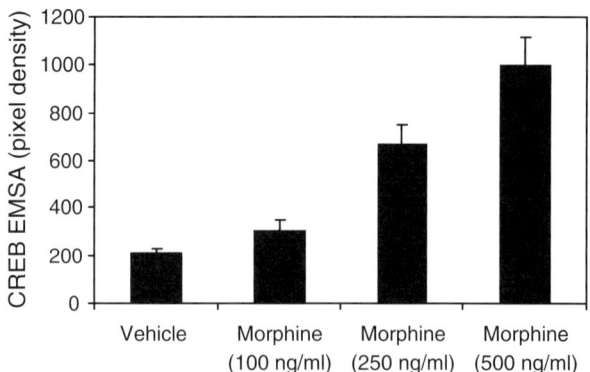

FIGURE 3. Effect of chronic morphine treatment on transcription factor CREB using electromobility shift assay. Naïve splenocytes were pretreated with either vehicle or morphine (250 ng/ml) and stimulated with plate bound anti-CD3/CD28 for 72 hr. At the end of the incubation period, CD4 T cells were prepared and re-stimulated with plate bound anti-CD3/CD28 for 6 hr. Nuclear protein was prepared and EMSA performed.

phosphorylation of MAP kinases and of the transcription factor CREB in dorsal root ganglion neurons.[67] Morphine treatment of human CEMx174 lymphocytic cells resulted in increased expression of MAPK cascade proteins. Morphine enhanced the cellular levels of ERK1, ERK2, MEK1, and MEKK in human lymphocytes through a naloxone sensitive receptor.[68] In a model of angiogenesis and tumor growth, morphine stimulated the MAPK/ERK signaling pathway.[69] In COS-7 cells transiently co-transfected with μ-, Δ-, or κ-opioid receptors and with ERK1- or ERK2-containing plasmids, opioid agonists have been shown to modulate the activity of the ERK. Pretreatment of cells with pertussis toxin abolished ERK1 and ERK2 activation by opioid agonists. Co-transfection of cells with the dominant negative mutant N17-Ras or with a $G\beta\gamma$ scavenger also suppressed opioid stimulation of ERK1 and ERK2.[70] The role of the MAPK signaling pathway in morphine-mediated CD4+ T-cell differentiation will thus be another important area of future investigation.

7. CLINICAL IMPLICATION OF MORPHINE-INDUCED Th1/Th2 DIFFERENTIATION

The existence of CD4+ T cells was strongly implied in early clinical observations. Leprosy was well known to exist in both healing (tuberculoid) and uncontrolled forms. The healing form of leprosy was associated with effective delayed type hypersensitivity (DTH) and low levels of antibody. The uncontrolled form was associated with high antibody titers and weak DTH.[71] The difficulty in interpretation emerged when it was discovered that both antibody and DTH reactions were mediated by CD4+ T cell (T-helper cells). The explanation arrived with the demonstration of T-cell subsets by Mosmann *et al.*[72] These investigators show that subsets of CD4+ T cells could be classified by patterns of cytokine production. The implication of CD4+ T-cell differentiation may be seen in a wide number of clinical states involving infectious disease, allergic diseases, and self-tolerance/autoimmunity.

Extracellular pathogens, especially parasitic helminthes, elicit Th2-dominated host responses. Th2-dependent IgG antibodies neutralize toxins produced by extracellular bacteria. It is controversial whether Th2-mediated IgE, eosinophil, and mast cell responses are protective.[19] Independent of the possible protective value of Th2 responses in helminthic infections, it is clear that Th2-host responses contribute to granuloma formation and hypereosinophilia.

Effective resistance to intracellular microbes including bacteria, protozoa, and fungi are associated with Th1 differentiation especially IFNγ and TNFα-mediated macrophage activation. Th1 responses are important in the host response to most viral infections. This response includes a wide set of effector mechanisms which include IFNα/β-activated NK cells, cytolytic CD8+ T cell (CTL), and antibodies with Th1 isotype pattern.

The classical pathway of IFNγ-dependent activation of macrophages by Th1-type responses is a well-established feature of cellular immunity to infection with intracellular pathogens, such as *M. tuberculosis*. An appropriate Th1 immune response is required for the elimination of *M. tuberculosis.*[73] In this regard, it has been demonstrated that a large increase in IL-4 and IL-13 synthesis correlates with lung damage. This is consistent with Th2 differentiation in the context of *M. tuberculosis* infection undermining the efficacy of immunity and contributing to immunopathology.[74] Given that Th2 cytokines inhibit Th1 responses, the question remains whether patients with prominent Th2 responses are more susceptible to *M. tuberculosis.*

The response to viral illness and the implication of Th2 differentiation is expressed in the host response to HIV infection. An early impairment in IL-2, IFNγ, and IL-12 production is observed in HIV-1 infection. It is thought that Th2-dominated responses play a pathogenic role in the host response to HIV and favor a more rapid evolution of HIV infection toward the full-blown disease.[75] It is controversial whether this alteration in cytokine production is the result of Th1 downregulation and viral progression. Of interest is the observation that HIV-1 preferentially infects Th2 clones.[19] This observation may explain the persistence of HIV virus in Th1 deficient hosts and may represent the link between opioid-induced Th2 differentiation, the course of HIV infection in opioid-addicted individuals.

8. SUMMARY AND CONCLUSION

Chronic morphine treatment has been shown to alter a number of immune parameters including suppression of cellular immunity. We speculate that differentiation of T helper to Th2 effector cells may be a major contributing factor to impaired cellular immunity following chronic morphine treatment. Our results show that chronic morphine treatment *in vitro* directs T-helper cells toward Th2 differentiation. We also show that chronic morphine treatment differentially modulates the transcriptional "switches" GATA3 and T-bet, thus providing a molecular mechanism by which morphine directs CD4+ differentiation. These studies suggest that therapies that prevent Th2 differentiation and promote Th1 cytokine synthesis may therefore prove beneficial in the immunosuppressed drug abuse population.

REFERENCES

1. Cantacuzene, J. (1898). Nouvelles recherches sur le monde de destruction des vibrions dans l'organisme. *Ann. Inst. Pasteur* **12:**273–300.
2. Brown, S. M., Stimmel, B., Taub, R. N., Kochwa, S., and Rosenfeld, R. E. (1974). Immunological dysfunction in heroin addicts. *Arch. Intern. Med.* **134:**1001–1006.
3. Louria, D. B., Hensle, T., and Rose, J. (1974). The major medical complications of heroin addiction. *Ann. Int. Med.* **67:**1–22.
4. McDonough. R. J., Madden. J. J., Falek. A., Shafer. D. A., Pline, M., Gordon. D. *et al* (1980). Alterations of T and null lymphocyte frequencies in the peripheral blood of human opiate addicts. *J. Immunol.* **125:**2539.
5. Wybran, J., Appelboon, T., Famey, J. P., and Govaerts, A. (1979). Suggestive evidence for receptors for morphine and methionine enkephalin on normal human blood T-lymphocytes. *J. Immunol.* **123:**1068–1070.

6. Roy, S. and Loh, H. H. (1996, November). Effects of opioids on the immune system. *Neurochem. Res.* **21**(11):1375–1386, review.

7. Peng, X., Mosser, D. M., Adler, M. W., Rogers, T. J., Meissler, J. J., Jr, and Eisenstein, T. K. (2000, November). Morphine enhances interleukin-12 and the production of other pro-inflammatory cytokines in mouse peritoneal macrophages. *J. Leukoc. Biol.* **68**(5):723–728.

8. Wang, J., Charboneau, R., Balasubramanian, S., Barke, R. A., Loh, H. H., and Roy, S. (2001, October). Morphine modulates lymph node-derived T lymphocyte function: Role of caspase-3, -8, and nitric oxide. *J. Leukoc. Biol.* **70**(4):527–536.

9. Limiroli, E., Gaspani, L., Panerai, A. E., and Sacerdote, P. (2002, July). Differential morphine tolerance development in the modulation of macrophage cytokine production in mice. *J. Leukoc. Biol.* **72**(1):43–48.

10. Guo, C. J., Li, Y., Tian, S., Wang, X., Douglas, S. D., and Ho, W. Z. (2002, November). Morphine enhances HIV infection of human blood mononuclear phagocytes through modulation of beta-chemokines and CCR5 receptor. *J. Investig. Med.* **50**(6):435–442.

11. Singhal, P. C., Kapasi, A. A., Reddy, K., Franki, N., Gibbons, N., and Ding, G. (1999, October). Morphine promotes apoptosis in Jurkat cells. *J. Leukoc. Biol.* **66**(4):650–658.

12. Singhal, P. C., Kapasi, A. A., Franki, N., and Reddy, K. (2000, May). Morphine-induced macrophage apoptosis: The role of transforming growth factor-beta. *Immunology* **100**(1):57–62.

13. Singhal, P. C., Bhaskaran, M., Patel, J., Patel, K., Kasinath, B. S., Duraisamy, S. *et al.* (2002, April 15). Role of p38 mitogen-activated protein kinase phosphorylation and Fas–Fas ligand interaction in morphine-induced macrophage apoptosis. *J. Immunol.* **168**(8):4025–4033.

14. Singhal, P., Kapasi, A., Reddy, K., and Franki, N. (2001). Opiates promote T cell apoptosis through JNK and caspase pathway. *Adv. Exp. Med. Biol.* **493**:127–135.

15. Roy, S., Balasubramanian, S., Sumandeep, S., Charboneau, R., Wang, J., Melnyk, D. *et al.* (2001, August). Morphine directs T cells toward T(H2) differentiation. *Surgery* **130**(2):304–309.

16. Peterson, P. K., Gekker, G., Brummitt, C., Pentel, P., Bullock, M., Simpson, M. *et al.* (1989, March). Suppression of human peripheral blood mononuclear cell function by methadone and morphine. *J. Infect. Dis.* **159**(3):480–487.

17. Suzuki, S., Chuang, L. F., Yau, P., Doi, R. H, and Chuang, R. Y. (2002, November 1). Interactions of opioid and chemokine receptors: Oligomerization of mu, kappa, and delta with CCR5 on immune cells. *Exp. Cell Res.* **280**(2):192–200.

18. Suzuki, S., Miyagi, T., Chuang, T. K., Chuang, L. F., Doi, R. H., and Chuang, R. Y. (2000, December 20). Morphine upregulates mu opioid receptors of human and monkey lymphocytes. *Biochem. Biophys. Res. Commun.* **279**(2):621–628.

19. Abbas, A., Murphy, K. M., and Sher, A. (1996). Functional diversity of helper T lymphocytes. *Nature* **383**:787–793.

20. Grogan, J. L. and Locksley, R. M. (2002). T helper cell differentiation: On again, off again. *Curr. Opin. Immunol.* **14**:366–372.

21. O'Connor, P. G., Shi, J. M., Henry, S., Durante, A. J., Friedman, L., and Selwyn, P. A. (1999, July). Tuberculosis chemoprophylaxis using a liquid isoniazid-methadone admixture for drug users in methadone maintenance. *Addiction* **94**(7):1071–1075.

22. Ray, A. and Cohn, L. (1999). Th2 cell and GATA3 in asthma: New insights into the regulation of airway inflammation. *J. Clin. Invest.* **104**(8):985–993.

23. Mullen, A. C., High, F. A., Hutchins, A. S. *et al.* (2001). Role of T-bet in commitment of Th1 cells before IL-12 dependent selection. *Science* **292**:1907–1910.

24. Robinson, D. S. and O'Garra, A. (2002). Further checkpoints in Th1 development. *Immunity* **16**:755–758.

25. Afkarian, M., Sedy, J. R., and Yang, J. *et al.* (2002, June). T-bet is a STAT1-induced regulator of IL-12R expression in naive CD4+ T cells [see comments]. Comment in: *Nat. Immunol.* **3**(6):506–508, 549–557.

26. Dong, C., and Flavell, R. A. (2002). Control of T helper cell differentiation—in search of master genes, *Science* 1–5.

27.   Flavell, R. A, Li, B., Dong, C., Lu, H. T., Yang, D. D., Enslen, H. *et al.* (1999). Molecular basis of T-cell differentiation. *Cold Spring Harbor Symposia on Quantitative Biology*, Vol LXIV, Cold Spring Harbor Laboratory Press, 0-87969-569-2/99, pp. 563–571.

28.   Losman, J., Chen, H., Jiang, P. *et al.* (1999). IL-4 signaling is regulated through the recruitment of phosphatases, kinases, and SOCS proteins to the receptor complex. *Cold Spring Harbor Symposia on Quantitative Biology*, Vol LXIV, Cold Spring Harbor Laboratory Press, 0-87969-569-2/99.

29.   O'Garra, A. and Arai, N. (2000). The molecular basis of T helper 1 and T helper 2 cell differentiation. *Trends Cell Biol.* **10:**542–550.

30.   Ranganath, S., Ouyang, W., Bhattarcharya, D., Sha, W. C., Grupe, A., Peltz, G. *et al.* (1998, October 15). GATA3-dependent enhancer activity in IL-4 gene regulation. *J. Immunol.* **161**(8):3822–3826.

31.   Szabo, S. J., Kim, S. T., Costa, G. L., Zhang, X., Fathman, C. G., and Glimcher, L. H. (2000). A novel transcription factor, T-bet, directs Th1 lineage commitment. *Cell* **100:**655–669.67. Zhang, D.H., Yang, L., and Ray, A. (1998). Differential responsiveness of the IL-5 and IL-5 genes to transcription factor GATA3. *J. Immunol.* **161:**3817.

32.   Arai, N., Lee, H. J., and Ferber, I. *et al.* (1999). Multiple levels of regulation of Th2 cytokine gene expression. *Cold Spring Harbor Symposia on Quantitative Biology*, Vol LXIV, Cold Spring Harbor Laboratory Press, 0-87969-569-2/99.

33.   Ho, I. C., Kim, J. I., Szabo, S. J., and Glimcher, L. H. (1999). Tissue specific regulation of cytokine gene expression. *Cold Spring Harbor Symposia on Quantitative Biology*, Vol LXIV, Cold Spring Harbor Laboratory Press, 0-87969-569-2/99, pp. 573–584.

34.   Ouyang, W., Lohning, M., Gao, Z., Assenmacher, M., Ranganath, S., Radbruch, A. *et al.* (2000, January). Stat6-independent GATA3 autoactivation directs IL-4-independent Th2 development and commitment. *Immunity* **12**(1):27–37.

35.   Ranganath, S. and Murphy, K. M. (2001, April). Structure and specificity of GATA proteins in Th2 development. *Mol. Cell. Biol.* **21**(8):2716–2725.

36.   Takeda, K., Tanaka, T., Shi, W., Matsumoto, M., Minami, M., Kashiwamura, S.-I. *et al.* (1996). Essential role of Stat6 in IL-4 signaling. *Nature* **380:**627–630.

37.   Dent, A. L., Shaffer, A. L., Yu, X., Allman, D., and Staudt, L. M. (1997). Control of inflammation, cytokine expression, and germinal center formation by BCL-6. *Science* **276:**589–592.

38.   Kaplan, M. H. and Grusby, M. J. (1998, July). Regulation of T helper cell differentiation by STAT molecules. *J. Leukoc. Biol.* **64**(1):2–5.

39.   Muraille, E. and Leo, O. (1998). Revisiting the Th1/Th2 paradigm. *Scand. J. Immunol.* **47:**1–9.

40.   Gorelik, L., Fields, P. *et al.* (2000). Cutting edge: TGF-b inhibits Th type 2 development through inhibition of GATA3 expression. *J. Immunol.* **165:**4773–4777.

41.   Kurata, H., Lee, H. J., O'Garra, A., and Arai, N. (1999). Ectopic expression of activated stat6 induces the expression of Th2 specific cytokines and transcription factors in developing Th1 cells. *Immunity* **11:**677–688.

42.   Kraus, J., Borner, C., Giannini, E., Hickfang, K., Braun, H., Mayer, P. *et al.* (2001). Regulation of mu-opioid receptor gene transcription by interleukin-4 and influence of an allelic variation within a STAT6 transcription factor binding site. *J. Biol. Chem.* 23;**276**(47):43901–43908.

43.   Sacerdote, P., San Secondo, VEMR. Sirchia, G., Manfredi, B., and Panerai, A. E. (1998). Endogenous opioids modulate allograft rejection time in mice: Possible relation with Th1/Th2 cytokines. *Clin. Exp. Immunol.* **113:**465–469.

44.   Sacerdote, P., Bianchi, M., and Panerai, A. E. (1996). *Regul. Pept.* **63:**79.

45.   Moynihan, J. A., Karp, J. D., Cohen, N., and Ader, R., (2000, January 24). Immune deviation following stress odor exposure: Role of endogenous opioids. *J. Neuroimmunol.* **102**(2):145–153.

46.   Berkenbosch, F., Vermes, I., and Tilders, F. J. (1984, September). The beta-adrenoceptor-blocking drug propranolol prevents secretion of immunoreactive beta-endorphin and alpha-melanocyte-stimulating hormone in response to certain stress stimuli. *Endocrinology* **115**(3):1051–1059.

47. Nelson, C. J., Carrigan, K. A., and Lysle, D. T. (2000). Naltrexone administration attenuates surgery-induced immune alterations in rats. *J. Surg. Res.* **94**(2):172–177.

48. Yin, D., Tuthill, D., Mufson, R. A., and Shi, Y. (2000, April 17). Chronic restraint stress promotes lymphocyte apoptosis by modulating CD95 expression. *J. Exp. Med.* **191**(8): 1423–1428.

49. Van den Bergh, P., Rozing, J., and Nagelkerken, L. (1994, April). Identification of distinct sites of beta-endorphin that stimulate lymphokine production by murine CD4+ T cells. *Lymphokine Cytokine Res.* **13**(2):63–69.

50. Van den Bergh, P., Dobber, R., Ramlal, S., Rozing, J., and Nagelkerken L. (1994, March). Role of opioid peptides in the regulation of cytokine production by murine CD4+ T cells. *Cell. Immunol.* **154**(1):109–122.

51. Van den Bergh, P., Rozing, J., and Nagelkerken, L. (1993, June). Beta-endorphin stimulates Ia expression on mouse B cells by inducing interleukin-4 secretion by CD4+ T cells. *Cell. Immunol.* **149**(1):180–192.

52. Roy, S., Balasubramanian, S., Sumandeep, S., Charboneau, R., Wang, J., Melnyk, D. *et al.* (2001, August). Morphine directs T cells toward Th2 differentiation. *Surgery* **130**(2):304–309.

53. Roy, S., Charboneau, R., and Barke, R. A. (1999). Morphine synergizes with LPS in an experimental sepsis model. *J.Neuroimmunol.* **95**:107–114.

54. Chen, C. H., Zhang, D. H., LaPorte, J. M., and Ray, A. (2000). Cyclic AMP activates p38 mitogen-activated protein kinase in Th2 cells: Phosphorylation of GATA3 and stimulation of Th2 cytokine gene expression. *J. Immunol.* **165**(10):5597–5605.

55. Kowalski, J. (1998, August). Immunomodulatory action of class mu-, delta- and kappa-opioid receptor agonists in mice. *Neuropeptides* **32**(4):301–306.

56. Schulz, S. and Hollt, V. (1998, March). Opioid withdrawal activates MAP kinase in locus coeruleus neurons in morphine-dependent rats in vivo. *Eur. J. Neurosci.* **10**(3):1196–1201.

57. Sharma, S. K., Klee, W. A., Nirenberg, M. (1975). Dual regulation of adenylate cyclase accounts for narcotic dependence and tolerance. *Proc. Natl. Acad. Sci. USA* **72**:3092–3096.

58. Sharma, S. K., Nirenberg, M., Klee, WA. (1975). Dual regulation of adenylate cyclase accounts for narcotic dependence and tolerance. *Proc. Natl. Acad. Sci. USA* **72**:590–594.

59. Lane-Ladd, S. B. *et al.* (1997). CREB in the locus coeruleus: Biochemical, physiological, and behavior evidence for a role in opiate dependence. *J. Neurosci.* **17**:7890–7901.

60. Nestler, E. J. and Aghajanian, G. K. (1997). Molecular and cellular basis of addiction. *Science* **278**(5335):58.

61. Avidor-Reiss, T., Nevo, I., Levy, R., Pfeuffer, T., and Vogel, Z. (1996, August 30). Chronic opioid treatment induces adenyl cyclase V superactivation: Involvement of Gbg. *J. Biol. Chem.* **271**(35):21309–21315.

62. Avidor-Reiss, T., Bayewitch, M., Levy, R., Matus-Leibovitch, N., Nevo, I., and Vogel, Z. (1995, December 15). Adenylcyclase super-sensitization in mu-opioid receptor-transfected Chinese hamster ovary cells following chronic opioid treatment. *J. Biol. Chem.* **270**(50):29732–29738.

63. Novak, T. J. and Rothenberg, E. V. (1990, December). cAMP inhibits induction of interleukin 2 but not of interleukin 4 in T cells. *Proc. Natl. Acad. Sci. USA* **87**(23):9353–9357.

64. Lee, H. J., Takemoto, N., Kurata, H., Kamogawa, Y., Miyatake, S., O'Garra, A., *et al.* (2000, July 3). GATA3 induces T helper cell type 2 (Th2) cytokine expression and chromatin remodeling in committed Th1 cells. *J. Exp. Med.* **192**(1):105–115.

65. Berhow, M. T., Hiroi, N., and Nestler, E. J. (1996, August 1). Regulation of ERK (extracellular signal regulated kinase), part of the neurotrophin signal transduction cascade, in the rat mesolimbic dopamine system by chronic exposure to morphine or cocaine. *J. Neurosci.* **16**(15):4707–4715.

66. Ortiz, J., Harris, H. W., Guitart, X., Terwilliger, R. Z., Haycock, J.W., and Nestler, E. J. (1995). Extracellular signal-regulated protein kinases (ERKs) and ERK kinase (MEK) in brain: Regional distribution and regulation by chronic morphine. *J. Neurosci.* **15**(2): 1285–1297.

67. Ma, W., Zheng, W. H., Powell, K., Jhamandas, K., and Quirion, R. (2001, October). Chronic morphine exposure increases the phosphorylation of MAP kinases and the transcription factor CREB in dorsal root ganglion neurons: An in vitro and in vivo study. *Eur. J. Neurosci.* **14**(7):1091–1094.

68. Chuang, L., Killian, K., and Chuang, R. Y. (1997). Induction of and activation of mitogen activated protein kinases of human lymphocytes as one of the signaling pathways of the immunomodulatory effects of morphine. *J. Biol. Chem.* **272**(43): 26815–26817.

69. Gupta, K., Kshirsagar, S., Liming Chang, Robert Schwartz, Ping, Y. Law, Doug Yee *et al.* (2002, August 1). Morphine stimulates angiogenesis by activating proangiogenic and survival-promoting signaling and promotes breast tumor growth. *Cancer Res.* **62**, 4491–4498.

70. Belcheva, M. M., Vogel, Z., Ignatova, E., and Avidum-Reiss, T. (1998). Opioid modulation of extracellular signal-regulated protein kinase activity is ras-dependent and involves Gbetagamma subunits. *J. Neurochem.* **70**(2):635–645.

71. Turk, J. L. and Bryceson, A. D. (1971). Immunological phenomena in leprosy and related diseases. *Adv. Immunol.* **171**(13): 209–266.

72. Mosmann, T. R., Cherwinski, H., Bond, M. W. *et al.* (1986). Two types of murine helper T cell clone. I. Definition according to profiles of lymphokine activities and secreted proteins. *J. Immunol.* **136**:2348–2357.

73. Beyers, A. D., van Rie, A., Adams, J., Fenhalls, G., Gie, R., and Beyers, N. (1998). Signals that regulate the host response to *Mycobacterium tuberculosis. Novartis Found. Symp.* **217**:145–157; discussion 157–159.

74. Romagnani, S. (2000, July). T-cell subsets (Th1 versus Th2). *Ann. Allergy Asthma Immunol.* **85**(1):9–18.

75. Rook, G. A. and Zumla, A. (2001, May). Advances in the immunopathogenesis of pulmonary tuberculosis. *Curr. Opin. Pulm. Med.* **7**(3):116–123.

11

Immunofluorescence Detection of Anti-CD3-ε-Induced Delta Opioid Receptors by Murine Splenic T Cells

BURT M. SHARP, KATHY McALLEN,
and NAHID A. SHAHABI

1. INTRODUCTION

Acting through opioid receptors, opiate alkyloids, and opioid peptides exert pleiotropic effects on cells involved in host defense and immunity.[1] These compounds are immunomodulators, modifying the immune response to mitogens, antigens, and antibodies that crosslink the T-cell receptor (TCR). More recent studies indicate that, by activating lymphocyte opioid receptors, these agents also can directly affect intracellular T-cell signaling and the function of other membrane receptors.[2–5]

Recently, it has become clear that immune cells express the same three mRNAs that encode the opioid receptor subtypes originally characterized in neuronal tissues.[6–13] Studies have used immunofluorescence and indirect fluorescence to demonstrate the regulated expression of both delta- and

BURT M. SHARP, KATHY McALLEN, and NAHID A. SHAHABI • Department of Pharmacology, University of Tennessee Health Science Center, Memphis, TN 38120.

Infectious Diseases and Substance Abuse, edited by Herman Friedman *et al.*
Springer, New York, 2005.

kappa-opioid receptors (i.e., DORs and KORs) on T cells and other immuno-cytes.[2,14,15] A relatively small fraction of the T cells in freshly obtained murine splenocytes and human peripheral blood mononuclear cells (PBMCs) have detectable DORs. However, marked increases in DOR protein have been reported after stimulation with staphylococcal enterotoxin B (SEB) *in vivo* and mitogens *in vitro.* [2,14] The studies reported herein were performed to character-ize the time-dependent expression of DORs by anti-CD3-ε-stimulated T cells in mixed splenocyte cultures. Using immunofluorescence flow cytometry, T-cell subsets expressing DOR were identified.

2. MATERIALS AND METHODS

2.1. Animals

Specific pathogen-free (SPF) 4–6-week-old female Balb/c mice were pur-chased from NCI (Bethesda, MD). They were maintained in an SPF facility on a 12 hr light/12 hr dark cycle, at a constant temperature (20°C), and allowed access *ad libitum* to food and water. All procedures were conducted in accordance with NIH guidelines for the care and use of laboratory animals as approved by the Animal Care and Use Committee of the Health Science Center, University of Tennessee.

2.2. Cell Preparation

Spleen cells were dispersed through a wire mesh, and red cells were lysed with ACK buffer (0.15M NH_4Cl, 1.0M $KHCO_3$, 0.01M Na-EDTA, pH 7.4). Cells were layered on Ficoll-Hypaque and centrifuged at 200× g for 7 min. The interface layer was washed in Hank's buffered saline without Ca^{2+} or Mg^{2+}, containing 2 mM EDTA and 0.1% gelatin. After centrifugation, cells were resuspended in RPMI 1640 containing penicillin (100 U/ml), streptomycin (50 μg/ml), 2 mM EDTA, 2 mM L-glutamine, 5×10^{-5} M 2-mercaptoethanol, and 5% fetal bovine serum. Small flasks were coated with anti-CD3-ε monoclonal antibody for 3 hr at room temperature. To remove excess mAb, flasks were washed three times with cold phosphate buffered saline. Splenocytes were cultured in ± precoated flasks for 24, 48, 72, or 120 hr.

2.3. Immunofluorescence Flow Cytometry

Splenocytes were fixed with 4% paraformaldehyde for 10 min at 4°C. Cells were washed with Tris-buffered saline (TBS) containing 1% donkey serum. They were then incubated with blocking buffer (5% donkey serum in TBS) overnight at 4°C. Cells were incubated with PE-anti-mouse-CD3-ε, PE-anti-mouse-CD4, or PE-anti-mouse-CD8 (PharMingen/Becton Dickinson Co., San Diego, CA) and anti-DOR antisera (1/400 dilution; Chemicon International, Inc., Temecula,

CA) for 2 hr at room temperature. Then they were washed, incubated with biotinylated donkey anti-rabbit IgG (Chemicon International, Inc., Temecula, CA) for 60 min, washed and incubated with fluorescein avidin DCS (Vector Labs, Burlingame, CA) for 10 min at 4°C. Thereafter, cells were washed three times with cold buffer, and cytofluorometric analyses were performed using an EPICS XL flow cytometer (Coulter, Miami, FL) equipped with an argon laser, and filtered for excitation at 488 nm and emission at 526 and 575 nm. For the background control, normal rabbit serum substituted for the primary antisera against DOR, and PE-rat-IgG2a,K (PharMingen/Becton Dickinson Co.) was used as isotype control.

3.   RESULTS

Figure 1 shows the time course of splenocyte DOR expression in response to plate-bound anti-CD3-ε. Without cross-linking, approximately 7–10% of splenocytes were DOR⁺. Anti-CD3-ε induced maximal DOR expression at 48 hr, when approximately 45% of the total splenocyte population was positive. Thereafter, a progressive decline was observed, although DOR was still present on twice as many anti-CD3-ε−stimulated cells after 120 hr in culture.

The second set of experiments evaluated the time-dependent expression of DOR by T cells (figure 2). In unstimulated cultures, approximately 2–5% of splenocytes were DOR⁺ T cells. Anti-CD3-ε induced DOR on T cells at all time intervals. This was maximal between 48and72 hr, at which time 25–30% of splenocytes were DOR⁺ T cells.

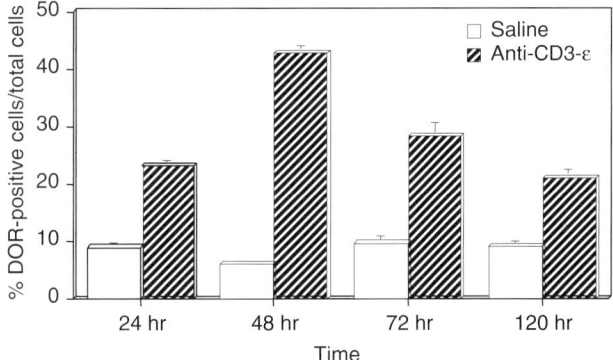

FIGURE 1. Immunofluorescence detection of DOR on murine splenocytes. Splenocytes were cultured with saline or immobile murine anti-CD3-ε for 24, 48, 72, and 120 hr. Cells were labeled with rabbit anti-DOR and normal rabbit serum (NRS) was used as a control. Anti-DOR was detected with a fluorescein avidin-biotin anti-rabbit Ab complex, and fluorescence levels detected in the presence of NRS/isotype were subtracted from the immunofluorescence signal emitted by the anti-DOR antibody. Data are expressed as the mean percentage of total splenocytes that were positive for DOR. Each column represents the mean ± SEM of three experiments, each in duplicate.

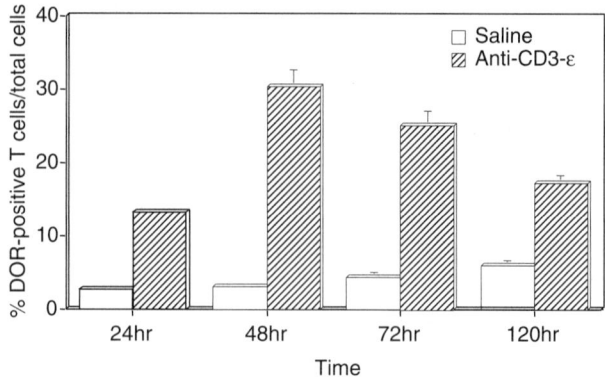

FIGURE 2. Immunofluorescence detection of DOR on murine T cells. Splenocytes were cultured with saline or anti-CD3-ε for 24, 48, 72, and 120 hr, and cytofluorometric analyses were performed as described in Fig.1. Data are shown as the percentage (mean ± sem; $n = 3$) of total splenocytes positive for both the T-cell marker, CD3-ε, and DOR (double positive). Each experiment was performed in duplicate.

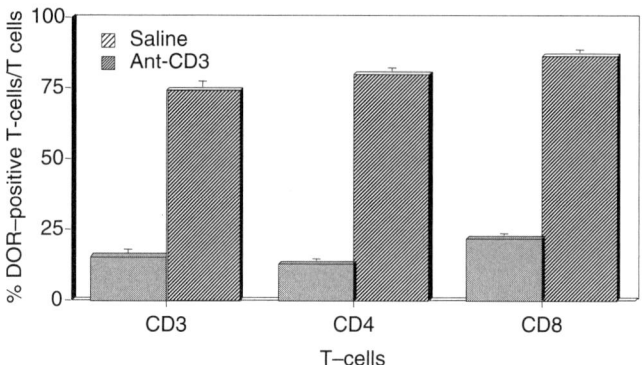

FIGURE 3. Immunofluorescence detection of DOR on total T cells compared to CD4[+] and CD8[+] subsets. Splenocytes were cultured with saline or anti-CD3-ε for 48 hr, and then labeled with rabbit anti-DOR and either anti-CD3-ε, anti-CD4 or anti-CD8, as described in the methods. Data are shown as the percentage (mean ± sem; $n = 3$) of cells positive for the following: both CD3-ε and DOR in the T-cell population; DOR and CD4 in the CD4[+] subset; and DOR and CD8 in the CD8[+] subset. Each experiment was performed in duplicate.

Figure 3 shows the expression of DOR by the T-cell subsets present in splenocytes cultured with anti-CD3-ε for 48 hr. TCR crosslinking induced DOR expression by 75% of all T cells compared to approximately 12% in controls. A similar fraction (i.e., 75%) of both CD4[+] and CD8[+] subsets were positive for DOR. In addition, anti-CD3-ε increased DOR immunofluorescence per cell by approximately 2-fold in the T-cell fraction and in both CD4[+] and CD8[+] subsets (1.90 ± 0.3-fold increase in T cells; 1.80 ± 0.2-fold in CD4[+]; and 1.93 ± 0.06-fold in CD8[+]).

4. DISCUSSION

These experiments demonstrate that anti-CD3-ε induced DOR expression on both CD4$^+$ and CD8$^+$ murine splenic T-cell subsets, and also increased the magnitude of DOR immunofluorescence per cell. The frequency of DOR expression by total T cells and CD4$^+$ and CD8$^+$ subsets was increased to 75% of each population, a 5–6-fold increase that was evident within 48–72 hr of activation. Maximal DOR expression occurs at a time interval when T cells are known to proliferate actively in response to anti-CD3-ε. Moreover, at this time interval, cells may be in the very early stages of differentiation into specific effectors (i.e., Th1 vs Th2). Thus, enhanced DOR expression may make a large fraction of T cells susceptible to endogenous and exogenous opioids that exert immunomodulatory effects on T-cell proliferation and differentiation.

Immunfluorescence microscopy and polymerase chain reaction (RT-PCR) were previously utilized to characterize DOR protein and mRNA expression *in vivo*, following a single injection of the superantigen, SEB.[2] Staphylococcal enterotoxins are known to provide a strong, specific antigenic stimulus that resembles the normal T-cell response to nominal antigen. [16,17] SEB binds to the MHC class II molecule and activates oligoclonal populations of T lymphocytes that express TCRs with homologous β chain variable regions (Vβ families).[18,19] SEB significantly enhanced murine splenocyte DOR mRNA expression 8 and 24 hr after injection. SEB also increased the fraction of the total splenocyte (5–20%) and T cell (8–50%) populations expressing DOR protein. In comparison to the present study, anti-CD3-ε induced DOR expression by a larger fraction of T cells. This is expected in view of the more restricted activation of T cells by SEB, which only affects 20–40% of all T cells.[20]

Immunfluorescence flow cytometry also was applied to detect DORs on subsets of human peripheral blood T cells after stimulating PBMCs with phytohemagglutinin (PHA) *in vitro*.[14] PHA stimulated the expression of DOR from basal levels of 2–20% of the PBMC population by 48 hr. DOR expression was approximately 40% of both the PHA-stimulated CD4$^+$ and CD8$^+$ T-cell subsets, and virtually all DORs were found on these subsets. Thus, anti-CD3-ε appeared to induce DOR on a greater fraction of murine splenic T-cell subsets compared to the effects of PHA on human peripheral blood T cells. This may reflect differences in species, cell compartment, and activating agent. Differential affinity of the anti-DOR antibody, which was generated against the murine DOR$_{3-17}$ sequence that differs by 8 out of 15 amino acids from the human homolog, is another factor that may potentially affect antibody binding and the measurement of murine vs human DOR levels.

In summary, mitogens, antigens, and TCR crosslinking all enhance T-cell expression of DOR. Based on studies *in vitro*, DOR is expressed by both CD4$^+$ and CD8$^+$ T cells. The effects of DOR ligands on T-cell signaling cascades (e.g., ERK 1, 2; JNK/ATF2), which have recently been described, suggest that DORs exert both direct and indirect effects on these pathways. Thus, DORs can attenuate TCR-dependent signaling[2] through ERK 1 and 2 and directly stimulate the phosphorylation of ATF2 through a mechanism that appears to depend on JNK (unpublished data).

ACKNOWLEDGEMENT. This work was supported by U.S.P.H.S. DA-04196.

REFERENCES

1. Sharp, B. M., Roy, S., and Bidlack, J. M. (1998). Evidence for opioid receptors on cells involved in host defense and the immune system. *J. Neuroimmunol.* **83**:45–56.
2. Shahabi, N. A., McAllen, K., Matta, S. G., and Sharp, B. M. (2000). Expression of delta opioid receptors by splenocytes from SEB-treated mice and effects on phosphorylation of MAP kinase. *Cell. Immunol.* **205**:84–93.
3. Shahabi, N. A., Daaka, Y., McAllen, K., and Sharp, B. M. (1999). Delta opioid receptors expressed by stably transfected Jurkat cells signal through the map kinase pathway in a ras-independent manner. *J. Neuroimmunol.* **94**:48–57.
4. Hedin, K. E., Bell, M. P., Huntoon, C. J., Karnitz, L. M., and McKean, D. J. (1999). Gi proteins use a novel βγ-and ras-independent pathway to activate extracellular signal-regulated kinase and mobilize AP-1 transcription factors in jurkat Tl lymphocytes. *J. Biol. Chem.* **274**:19992–20001.
5. Rogers, T. J., Steele, A. D., Howard, O. M., and Oppenheim, J. J. (2000). Bidirectional heterologous desensitization of opioid and chemokine receptors. *Ann. NY Acad. Sci.* **917**:19–28.
6. Chuang, L. F., Chuang, T. K., Killam, K. F., Chuang, A. J., Kung, H. F., Yu, L. *et al.* (1994). Delta opioid receptor gene expression in lymphocytes. *Biochem. Biophys. Res. Commun.* **202**:1291–1299.
7. Sharp, B. M., Shahabi, N. A., McKean, D., Li MD, and McAllen, K. (1997). Detection of basal levels and induction of delta opioid receptor mRNA in murine splenocytes. *J. Neuroimmunol.* **78**:198–202.
8. Li MD, McAllen, K., and Sharp, B. M. (1999). Regulation of delta opioid receptor expression by anti-CD3-ε, PMA, and ionomycin in murine splenocytes and T cells. *J. Leukoc. Biol.* **65**:707–714.
9. Miller, B. (1996). Delta opioid receptor expression is induced by concanavalin A in CD4+ T cells. *J. Immunol.* **157**:5324–5328.
10. Gaveriaux, C., Peluso, J., Simonin, F., Laforet, J., and Kieffer, B. (1995). Identification of kappa- and delta-opioid receptor transcripts in immune cells. *FEBS Lett.* **369**:272–276.
11. Chuang, L. F., Chuang, T. K., Killam, K. F., Qiu, O., Wang, X. R., Lin, J. J. *et al.* (1995). Expression of kappa opioid receptors in human and monkey lymphocytes. *Biochem. Biophys. Res. Commun.* **209**:1003–1010.
12. Sedqi, M., Roy, S., Ramakrishnan, S., Elde, R., and Loh, H. H. (1995). Complementary DNA cloning of a μ-opioid receptor from rat peritoneal macrophages. *Biochem. Biophys. Res. Commun.* **209**:563–574.
13. Chang, T. K., Killam, K. F., Jr., Chuang, L. F., Kung, H. F., Sheng, W. S., Chao, C. C. *et al* (1995). Mu opioid receptor gene expression in immune cells. *Biochem. Biophys. Res. Commun.* **216**:922–930.
14. Sharp, B. M., McAllen, K., Gekker, G., Shahabi, N. A., and Peterson, P. K. (2001). Immunofluorescence detection of δ opioid receptors (DOR) on human peripheral blood CD4$^+$ T cells and DOR-dependent suppression of HIV-1 expression. *J. Immunol.* **167**:1097–1102.
15. Lawrence, D. M. P., El-Hamouly, W., Archer, S., Leary, J. F., and Bidlack, J. M. (1995). Identification of κ opioid receptors in the immune system by indirect immunofluorescence. *Proc. Natl. Acad. Sci. USA* **92**:1062–1066.
16. Pinto, M., Torten, M., and Birnbaum, S. C. (1978). Suppression of the in vivo humoral and cellular immune response by staphylococcal enterotoxin B (SEB). *Transplantation* **25**:320–323.

17. Donnelly, R. P. and Rogers, T. J. (1982). Immunosuppression induced by staphylococcal enterotoxin B. *Cell Immunol.* **72**:166–177.

18. Herman, A., Kappler, J. W., Marrack, P., and Pullen, A. M. (1991). Superantigens: Mechanism of T-cell stimulation and role in immune responses. *Annu. Rev. Immunol.* **9**:745–772.

19. Callahan, J. E., Herman, A., Kappler, J. W., and Marrack, P. (1990). Stimulation of B10.BR T cells with superantigenic staphylococcal toxins. *J. Immunol.* **144**:2473–2479.

20. Marrack, P. and Kappler, J. (1990). The staphylococcal enterotoxins and their relatives. *Science* **248**:705–711.

12

Modulation of Immune Function by Novel Opioid Receptor Ligands

RICHARD WEBER and RICARDO GOMEZ

1. INTRODUCTION

Morphine and other natural alkaloid opiates have been used in medicine for centuries. Synthesis of analogs of opiate alkaloids and primary structure activity studies have an almost hundred year history. The endogenous opioid peptides, their genetic expression, and enzymatic metabolism have been described. A number of nonpeptide and peptide analogs have led to characterization of opioid receptor types (mu, delta, and kappa) and their subtypes, and very recently, all types of opioid receptors of different species have been characterized at the molecular level. The progressive study of the opioid system allowed the introduction of various new types of drugs. In addition, the opioid system is often used as a model for studies in neurobiology as well as in bioorganic chemistry.[1] It is now clear that opioid receptors participate in the function of the cells of the immune system, and evidence suggests that opioids modulate both innate and acquired immune responses. It is proven that μ-, κ-, and Δ-opioid compounds can alter resistance to a variety of infectious agents, including the human immunodeficiency virus (HIV) (reviewed in Gomez-Flores and Weber[2]), and augment cancer development, as reported in several studies showing an increase of metastasis in different models of tumor growth.[3,4]

RICHARD WEBER • Department of Biomedical and Therapeutic Sciences, University of Illinois College of Medicine at Peoria, Peoria, IL. RICARDO GOMEZ • Universidad Autónoma de Nuevo León, San Nicolás de los Garza, NL, México.

Infectious Diseases and Substance Abuse, edited by Herman Friedman *et al.*
Springer, New York, 2005.

Modulation of the inflammatory response appears to be a target of these compounds, including effects on lymphocyte and natural killer (NK) cell function, phagocytic activity, as well as the response of leukocytes to various chemokines (reviewed in Gomez-Flores and Weber,[2] Ishikawa et al.[3]). Moreover, findings from several laboratories have demonstrated the impact of opioid treatment on antibody responses, and the molecular basis for this effect is likely due, at least in part, to the modulation of both cytokine and cytokine receptor expression.[5]

2. MORPHINE-INDUCED IMMUNOREGULATION

Morphine, as a model for opioid-mediated immunoregulation, has been reported to have immunosuppressive effects following *in vivo* administration, which were observed to be mediated by μ-opioid receptors found within the central nervous system (CNS).[6–11] These observations have been extended in showing that central opioid receptor activation may or may not involve both the hypothalamic pituitary adrenal (HPA) axis[11,12] and adrenergic pathways[13–15] in suppressing NK activity following acute morphine administration. Other studies have also found that unidentified central pathways are involved in the immunosuppressive effects of morphine.[16,17] Taken together, these observations suggest that the administration of opiates (e.g., morphine or heroin) has no direct overall effect on immune cells but rather operates through indirect means involving predominantly central pathways that ultimately modify immune function.

The major effect of strong mu agonists *in vivo* is immunosuppressive, and indirect and direct *in vitro* effects of peptides and novel nonpeptides have been well substantiated.[18] Paradoxically, the direct effect of certain opioids on leukocytes can enhance, suppress, or have no effect on *in vitro* and *in vivo* parameters of immune function. Eisenstein *et al.* have previously demonstrated that opioids directly affect cellular and humoral immune functions through classical opioid receptors.[19,20] This research group has shown that μ-, κ-, and Δ-opioid receptors were associated with regulating lymphoid cell production of antibodies.[21,22] It is now clear that endogenous opioid peptides and exogenous opioid alkaloids modulate the immune function by directly acting on opioid receptors on the surface of cells involved in host defense and immunity. Opioid receptors expressed by immune cells are related to neuronal-type opioid receptors, particularly κ- and δ-opioid receptors.[23,24] Opioids may act like cytokines, and both types of molecules share many properties including paracrine, autocrine, and endocrine sites of action, functional redundancy, pleiotropy, and effects that are both dose and time dependent.[23]

3. NOVEL OPIOID DERIVATIVES

Although the CNS-mediated indirect effect of opioids has been shown to suppress immune function (see above), the direct effect of certain novel opioid derivatives on cells of the immune system have been shown by our group to induce immunopotentiation *in vitro*.[25–29] In this respect, in early studies, we have observed that the nonpeptide delta agonist (+)-4-[(alpha R)-alpha-((2S, 5R)-4-allyl-2,5-dimethyl-1-piperazinyl)-3-methoxybenzyl]-*N*,*N*-diethyl-benzamide (SNC80; Fig. 1) was capable of stimulating proliferation of rat thymic and splenic lymphocytes when costimulated with concanavalin A (Con A); however, SNC80 did not affect human lymphocyte proliferative response (Table I). SNC80 was also found to inhibit HIV replication in T lymphocytes by suppressing production of p24 antigen.[24,30] These *in vitro* findings suggested that the anti-HIV-1 property of SNC80 might have therapeutic potential for treating patients with acquired immunodeficiency syndrome.[24] In addition, we have reported that *in vitro* treatment with SNC80 significantly stimulated production of tumor necrosis factor-alpha (TNF-α) by resident and LPS-activated rat macrophages (Table I; Gomez-Flores and Weber[29]). SNC80 was also observed to stimulate TNF-α production by LPS-activated human macrophages (Table I), and increase TNF-α (marginal 1.15-fold increase at 10^{-7}M) and IL-8 (1.46-, 1.94-, and 1.19-fold increases at 10^{-7}M, 10^{-8}M, and 10^{-9}M, respectively) mRNA signal in human peripheral blood mononuclear cells (unpublished observations). Furthermore, SNC80 was recently reported to increase rat and human leukocyte chemotaxis (Table I).[31] In this respect, it has been recognized that μ-, Δ-, and κ-opioid agonists are capable of stimulating T-cell chemotaxis.[32]

We have also evaluated the *in vivo* and *ex vivo* effects of SNC80 on immune functions and tumor growth. We have reported that intravenous administration of SNC80 (6.8 mg/kg) increased the production of TNF-α and nitric oxide (NO) by LPS-stimulated splenic macrophages (Table I). Intravenous injection of SNC80 plus Con A also potentiated LPS-stimulated macrophage functions *ex vivo*.[29] Furthermore, in an *in vivo* tumor model of opioid action, we have observed that administration of SNC80 significantly increased L5178Y-R tumor-bearing mice survival and reduced tumor weights (Table I) in these animals (unpublished observations). Because the direct effects of SNC80 on L5178Y-R cell line were marginal (data not shown), it is possible that SNC80 induced a proinflammatory state in these animals leading to tumor cell destruction. In this

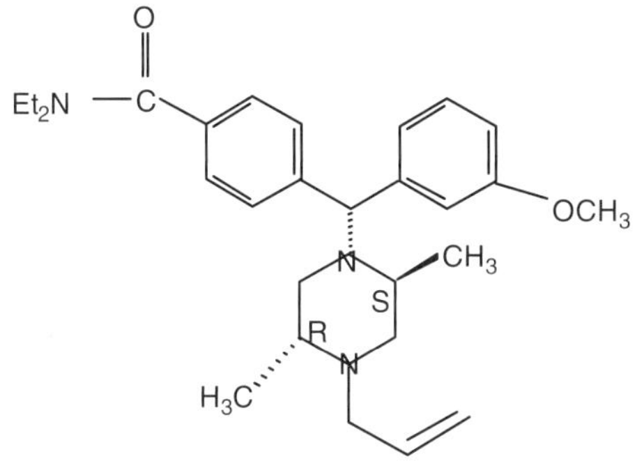

FIGURE 1. SNC 80 structure.

TABLE I
Immunological Alterations by Nonpeptide Opioids

	Lymphoproliferation		NO production		TNF-α production		Chemotaxis[a]		Tumor cell growth[b]
	Rat	Human	Rat	Human	Rat	Human	Rat	Human	(vs L5178Y-R)
SNC80	↑[b]	➡[b]	↑[c]	ND	↑[c]	↑[b]	↑	↑	↓
Morphinan[b]									
R = CH₃	↑	ND	➡	ND	➡	ND	ND	ND	↓
R = C₆H₅	↑↑↑	ND	➡	ND	➡	ND	ND	ND	↓
R = OH	↑	ND	➡	ND	➡	ND	ND	ND	↓
R = NH₂	↑	ND	➡	ND	➡	ND	ND	ND	↓
Pyrazol	↑↑	ND	➡	ND	➡	ND	ND	ND	↓
CGPM-9[d]	↑	ND	↓	ND	↓	ND	ND	ND	ND

[a]Unpublished observations.
[b]Gomez-Flores and Weber (2001).
[c]Ordaz-Sanchez et al. (2003).
[d]Hicks et al. (2001).
↑ increased; ↓ decreased; ➡ no effect; ND, not done.

respect, SNC80 was recently shown to increase chemotaxis of human and rat leukocytes (Table I), that then may be activated by this opioid at the site of the tumor, as suggested for our *in vitro* and *ex vivo* studies.[31] A prerequisite to the development of an efficient cell and/or gene therapy for certain types of cancers is a precise characterization of the inflammatory cell populations present in the tumor stroma associated with a cancer. In general, tumor stroma inflammatory cells might be mainly tumor-infiltrating lymphocytes (TIL) (approximately two thirds) (among them, 80% are T cells) and tumor-associated macrophages (TAM) (approximately one third).[33] Activated macrophages secrete several substances such as TNF-α and NO that are directly involved in tumor cell death.[34] TNF-α is known to cause tissue inflammation, tumor cell death, and toxic side effects such as body weight reduction.[35] Similarly, NO is the main reactive nitrogen intermediate (RNI) with a diverse range of actions in both physiological and pathological processes. Its role in tumor biology remains unknown. NO is known to have both tumor promoting and inhibitory effects, presumed to be dependent on its local concentration within the tumor.[36] On the other hand, SNC80-mediated activation of lymphocytes may lead to the release of cytokines such as interferon-gamma and IL-2 that are involved in inflammatory processes.[37] Recruitment and activation of macrophages and lymphocytes may significantly account for the increased survival of tumor-bearing mice and the reduction in tumor mass observed after *in vivo* SNC80 treatment. These data demonstrate that SNC80 may be a potent inducer of adaptive immune responses against tumor cells and may represent a potentially useful tool in the immunotherapy of certain types of cancers. In addition, SNC80 may act as an immunological adjuvant in a vaccine regime that may improve antitumor immunity by stimulating the induction of Th1-promoting cytokines. There is also increasing evidence that many adjuvants induce Th1-type cytokines, which correlates with the induction of antitumor immunity.[38] Th1-type responses that comprise cell-mediated immunity are characterized by the secretion of interferon-gamma by T cells that are induced by antigen-presenting cell-derived IL-12.[38] Therefore, the use of Th1-inducing adjuvants may provide an essential strategy for the future success of immunotherapy.

4. OPIOID RECEPTORS AND ANALGESIA

Stimulation of μ-opioid receptors has the potential to relieve pain[39]; but in addition to analgesic properties, μ-opioid receptor agonists have been associated with immunosuppression through central or peripheral pathways[7,40,41] (reviewed in Gomez-Flores and Weber[9]). We have utilized novel μ-opioid receptor selective agonists that are not immunosuppressive, and some of which are strong immunopotentiating agents. We evaluated the effects of morphinans with substituted pyrimidine (methyl, phenyl, hydroxyl, and amino groups) and pyrazole groups (Fig. 2), on *in vitro* rat thymocyte proliferation, splenic macrophage functions, and *in vitro* tumor cell growth. The μ-opioid receptor selective morphinans, such as levallorphan, cyclorphan, and butorphanol, are oxymorphon derivatives that were first introduced by Grewe in 1946. They are similar in structure to the morphine analogs, but lack the E ring found in the naturally occurring opioids, as well as the 6-OH and the 7,8-double bond. In our studies, we observed that morphinans at concentrations of 10^{-10}–10^{-5} M increased T-lymphocyte proliferation with the order of potency phenyl > pyrazol > hydroxyl, amino, methyl > Con A alone (control) (in preparation). These results indicated that the inclusion of a phenyl substituent at the 2′ position of the pyrimidine group significantly potentiated lymphoproliferation. In contrast, no alterations in macrophage functions were observed, suggesting a selective effect on lymphocytes (unpublished observations). Differential effects of opioids on leukocyte functions are commonly observed. In this regard, Kowalski *et al.*[42] reported that enkephalins were associated with both suppressing and enhancing effects on splenic NK cell and macrophage functions related to the treatment period. In addition, Pacifici *et al.*[43] observed time-dependent biphasic effects of

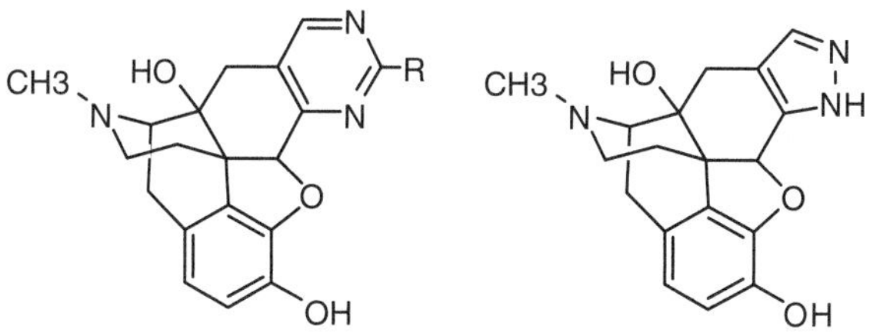

1a, R = CH₃
1b, R = C₆H₅
1c, R = OH
1d, R = NH₂

FIGURE 2. Morphinan derivatives structure.

morphine, but not methadone, on immune parameters *in vivo*. In yet another study, dose-dependent bimodal responses of lymphocytes and macrophages to opioids have been reported.[44,45]

Ryng *et al.*[46] also demonstrated that non-opioid substituted phenilamides of 5-amino-3-methylisooxazole-4-carboxylic acid have differential effects on lymphocyte and macrophage functions, and Hicks *et al.*[28] reported that the tetrahydroquinoline CGPM-9 enhanced rat thymic lymphoproliferation, but suppressed NO and TNF-α production by peritoneal macrophages (see below). The mechanism(s) by which morphinans enhance lymphocyte proliferative response, but do not alter macrophage functions, remains to be investigated. However, potentiating lymphocyte functions while not activating macrophages may be advantageous for these opioid derivatives. Macrophage activation can cause both poditive and negative effects during the inflammation process.[47,48] Stimulation of lymphoproliferation by morphinan derivatives may be utilized in clinical situations where lymphocyte populations are significantly reduced, as in the cases of AIDS and aging.[49]

5. IMMUNOENHANCEMENT

We have also found that the opioid 4-tyrosylamido-6-benzyl-1,2,3,4 tetrahydroquinoline (CGPM-9) possesses immunoenhancing properties *in vitro*.[26,28] CGPM-9 is a high affinity ligand with moderate μ-opioid receptor selectivity.[50] We observed that CGPM-9 potentiated Con-A-induced thymic T-lymphocyte proliferation and suppressed peritoneal macrophage production of NO and TNF-α[28] (Table I). Divergent or opposite effects of opioids on leukocyte functions are commonly observed.[51,52] The mechanism(s) by which CGPM-9 enhances lymphocyte proliferative response but suppresses macrophage NO and TNF-α production remains to be elucidated. However, potentiating lymphocyte functions while suppressing those of macrophages may be advantageous for this opioid. NO and TNF-α are produced during inflammation and can be both beneficial and detrimental for the organism.[48,53] Although these molecules are usually associated with antimicrobial and antitumor activities,[53] they also induce immunosuppression by affecting lymphocyte and macrophage functions through direct action on these cells,[54,55] or indirectly via the CNS.[56,57] In addition, suppression of macrophage functions by CGPM-9 may be mediated by an autocrine mechanism involving the induction of IL-10 or TGF-β.[58] Therefore, CGPM-9 may potentiate lymphoproliferative responses with increased cytokine release, while suppressing macrophage functions and potential pathological states.

6. CONCLUSIONS AND DISCUSSION

Nonpeptide opioid agonists are known to be not only highly selective and potent, but also proteolytically stable, thus increasing their clinical applications.[59]

The clinical use of properly designed and synthesized opioid ligands could serve as immunotherapeutic agents with potential use in the treatment of diseases such as AIDS and cancer. In addition, because surgical stress also induces immune dysfunction, there is an urgent need to search for analgesic drugs devoid of immunosuppressive effects. It is clear that knowledge of how opioids produce direct effects on the immune system may allow the discovery, design, and synthesis of new opioids that have specific immunoregulatory properties, which could potentially be utilized in many different clinical situations where immunosuppression is undesirable, as shown for μ-selective ligands such as morphine.[60] Because of their effects on immune function, μ-opioid agonists might not be optimal for management of moderate to severe pain following a variety of surgical procedures, cancer, and other related traumatisms. However, opioid derivatives such as morphinans or SNC80 may have the potential to not only stimulate the immune system, but also the capacity to inhibit tumor cell growth, making these compounds potentially suitable to treat pain and enhance the immune status of immunocompromised individuals against cancer and infectious diseases.

REFERENCES

1. Misicka, A. (1995). Peptide and nonpeptide ligands for opioid receptors. *Acta Pol. Pharm.* **52**(5):349–363.
2. Gomez-Flores, R. and Weber, R. J. (1999a). Opioids, opioid receptors, and the immune system. In N. Plotnikoff, R. Faith, A. Murgo, and R. Good (eds), *Cytokines—Stress and Immunity*, CRC Press, Boca Ratón, FL, pp. 281–314.
3. Ishikawa, M., Tanno, K., Kamo, A., Takayanagi, Y., and Sasaki, K. (1993). Enhancement of tumor growth by morphine and its possible mechanism in mice. *Biol. Pharm. Bull.* **16**:762–766.
4. Yeager, M. P. and Colacchio, T. A. (1991). Effect of morphine on growth of metastatic colon cancer *in vivo. Arch Surg.* **126**(4):454–456.
5. McCarthy, L., Wetzel, M., Sliker, J. K., Eisenstein, T. K., and Rogers, T. J. (2001). Opioids, opioid receptors, and the immune response. *Drug Alcohol Depend.* **62**(2):111–123.
6. Shavit, Y., Depaulis, A., Martin, F. C., Terman, G. W., Pechnick, R. N., Zane, C. J. *et al.* (1986). Involvement of brain opiate receptors in the immune-suppressive effect of morphine. *Proc. Natl. Acad. Sci. USA* **83**:7114–7117.
7. Weber, R. J. and Pert, A. (1989). The periaquaductal gray matter mediates opiate-induced immunosuppression. *Science* **245**:188–190.
8. Gomez-Flores, R. and Weber, R. J. (2000). Differential effects of buprenorphine and morphine on immune and neuroendocrine functions following acute administration in the rat mesencephalon periaqueductal gray. *Immunopharmacology* **48**:145–156.
9. Gomez-Flores, R. and Weber, R. J. (1999b). Inhibition of IL-2 production and down regulation of IL-2 and transferrin receptors on rat splenic lymphocytes following PAG morphine administration: A role in NK and T cell suppression. *J. Interferon Cytokine Res.* **19**:625–630.
10. Gomez-Flores, R., Jin-Liang, S., and Weber, R. J. (1999c). Suppression of splenic macrophage functions after acute morphine action in the rat mesencephalon periaqueductal gray. *Brain Behav. Immun.* **13**:212–224.
11. Suo, J. L., Gomez-Flores, R., and Weber, R. J. (2002). Immunosuppression induced by central action of morphine is not blocked by mifepristone (RU 486). *Life Sci.* **71**(22):2595–2602.

12. Freier, D. O. and Fuchs, B. A. (1993). Morphine-induces alterations in thymocyte subpopulations of B6C3F1 mice. *J. Pharmacol. Exp. Ther.* **265**:81–88.

13. Carr, D. J. J., Mayo, S., Gebhardt, B. M., and Porter, J. (1994). Central α-adrenergic involvement in morphine-mediated suppression of splenic NK activity. *J. Neuroimmunol.* **53**:53–63.

14. Carr, D. J. J., Gebhardt, B. M., and Paul, D. (1995). α-Adrenergic and μ2 opioid receptors are involved in morphine-induced suppression of splenocyte natural killer activity. *J. Pharmacol. Exp. Ther.* **264**:1179–1186.

15. Hall, D. M., Suo, J.-L., and Weber, R. J. (1998). Opioid mediated effects on the immune system: Sympathetic nervous system involvement. *J. Neuroimmunol.* **83**:29–35.

16. Flores, L. R., Hernandez, M. C., and Bayer, B. M. (1994). Acute immunosuppressive effects of morphine: Lack of involvement of pituitary and adrenal factors. *J. Pharmacol. Exp. Ther.* **268**:1129–1134.

17. Hernandez, M. C., Flores, L. R., and Bayer, B. M. (1993). Immunosuppression by morphine is mediated by central pathways. *J. Pharmacol. Exp. Ther.* **267**:1336–1341.

18. Weber, R. J. and Pert, A. (1984). Opiatergic modulation of the immune system. In E. Muller and A. Genazzani (eds), *Central and Peripheral Endorphins: Basic and Clinical Aspects*, New York, pp. 35–42.

19. Eisenstein, T. K., Meissler, J. J. Jr., Rogers, T. J., Geller, E. B., and Adler, M. W. (1995). Mouse strain differences in immunosuppression by opioids in vitro. *J. Pharmacol. Exp. Ther.* **275**(3):1484–1489.

20. Rojavin, M., Szabo, I., Bussiere, J. L., Rogers, T. J., Adler, M. W., and Eisenstein, T. K. (1993). Morphine treatment in vitro or in vivo decreases phagocytic functions of murine macrophages. *Life Sci.* **53**:997–1006.

21. Taub, D. D., Eisenstein, T. K., Geller, E. B., Adler, M. W., and Rogers, T. J. (1991). Immunomodulatory activity of mu- and kappa-selective opioid agonists. *Proc. Natl. Acad. Sci. USA* **88**(2):360–364.

22. Rahim, R. T., Meissler, J. J. Jr., Cowan, A., Rogers, T. J., Geller, E. B., Gaughan, J. *et al.* (2001, October). Administration of mu-, kappa- or delta2-receptor agonists via osmotic minipumps suppresses murine splenic antibody responses. *Int. Immunopharmacol.* **1**(11):2001–2009.

23. Bidlack, J. M. (2000). Detection and function of opioid receptors on cells from the immune system. *Clin. Diagn. Lab. Immunol.* **7**(5):19–23.

24. Sharp, B. M., Roy, S., and Bidlack, J. M. (1998). Evidence for opioid receptors on cells involved in host defense and the immune system. *J. Neuroimmunol.* **83**(1–2):45–56.

25. Sanchez, S. R., Calderon, S. N., Rice, K. C., Riley, M. E., and Weber, R. J. (1996). Receptor mediated stimulation of lymphocyte proliferation by novel δ-selective opioid ligands, College for the Problems of Drug Dependence, 58th Annual Scientific Meeting, San Juan, Puerto Rico.

26. Riley, M. E., Ananthan, S., and Weber, R. J. (1998). Novel non-peptidic opioid compounds with immunopotentiating effects. *Adv. Exp. Med. Biol.* **437**:183–187.

27. Nowak, J. E., Gomez-Flores, R., Calderon, S. N., Rice, K. C., and Weber, R. J. (1998). Rat NK cell, T cell, and macrophage functions following intracerebroventricular injection of SNC 80. *J. Pharmacol. Exp. Ther.* **286**:931–937.

28. Hicks, M. E., Gomez-Flores, R., Wang, C., Mosberg, H., and Weber, R. J. (2001). Differential effects of the novel non-peptidic opioid 4-tyrosylamido-6-benzyl-1,2,3,4 tetrahydroquinoline (CGPM-9) on in vitro T lymphocyte and macrophage functions. *Life Sci.* **68**:2685–2694.

29. Gomez-Flores, R. and Weber, R. J. (2001). Increased nitric oxide and TNF-α production by rat macrophages following in vitro stimulation and intravenous administration of SNC 80. *Life Sci.* **68**:2675–2684.

30. Sharp, B. M., McAllen, K., Gekker, G., Shahabi, N. A., and Peterson, P. K. (2001). Immunofluorescence detection of delta opioid receptors (DOR) on human peripheral blood CD4+ T cells and DOR-dependent suppression of HIV-1 expression. *J. Immunol.* **167**:1097–1102.

31. Ordaz-Sanchez, I., Weber, R. J., Rice, K. C., Zhang, X., Rodríguez-Padilla, C., Tamez-Guerra, R. *et al.* (2003). Chemotaxis of human and rat leukocytes by the delta-selective non-peptidic opioid SNC 80. *Rev. Latinoam. Microbiol.* **45**(1–2):16–23.

32. Rogers, T. J., Steele, A. D., Howard, O. M., and Oppenheim, J. J. (2000). Bidirectional heterologous desensitization of opioid and chemokine receptors. *Ann. N. Y. Acid. Sci.* **917**:19–28.

33. Kataki, A., Scheid, P., Piet, M., Marie, B., Martinet, N., Martinet, Y. *et al.* (2002). Tumor infiltrating lymphocytes and macrophages have a potential dual role in lung cancer by supporting both host-defense and tumor progression. *J. Lab. Clin. Med.* **140**:320–328.

34. Klimp, A. H., de Vries, E. G., Scherphof, G. L., and Daemen, T. (2002). A potential role of macrophage activation in the treatment of cancer. *Crit. Rev. Oncol. Hematol.* **44**:143–161.

35. Kamada, H., Tsutsumi, Y., Yamamoto, Y., Kihira, T., Kaneda, Y., Mu, Y. *et al.* (2000). Antitumor activity of tumor necrosis factor-alpha conjugated with polyvinylpyrrolidone on solid tumors in mice. *Cancer Res.* **60**:6416–6420.

36. Brennan, P. A. and Moncada, S. (2002). From pollutant gas to biological messenger: The diverse actions of nitric oxide in cancer. *Ann. R. Coll. Surg. Engl.* **84**:75–78.

37. Katsikis, P. D., Cohen, S. B., Londei, M., and Feldmann, M. (1995). Are CD4+ Th1 cells pro-inflammatory or anti-inflammatory? The ratio of IL-10 to IFN-gamma or IL-2 determines their function. *Int. Immunol.* 1995 **7**:1287–1294.

38. Dredge, K., Marriott, J. B., Todryk, S. M., and Dalgleish, A. G. (2002). Adjuvants and the promotion of Th1-type cytokines in tumour immunotherapy. *Cancer Immunol. Immunother.* **51**:521–531.

39. Pert, C. B. and Snyder, S. H. (1973). Properties of opiate-receptor binding in rat brain. *Proc. Natl. Acad. Sci. USA* **70**:2243–2247.

40. Carr, D. J. J., Gebhardt, B. M., and Paul, D. (1993). Alpha adrenergic and mu-2 opioid receptors are involved in morphine-induced suppression of splenocyte natural killer activity. *J. Pharmacol. Exp. Ther.* **264**:1179–1186.

41. Lysle, D. T., Hoffman, K. E., and Dykstra, L. A. (1996). Evidence for the involvement of the caudal region of the periaqueductal gray in a subset of morphine-induced alterations of immune status. *J. Pharmacol. Exp. Ther.* **277**:1533–1540.

42. Kowalski, J., Belowski, D., and Wielgus, J. (1995). Bidirectional modulation of mouse natural killer cell and macrophage cytotoxic activities by enkephalins. *Pol. J. Pharmacol.* **47**:327–331.

43. Pacifici, R., Patrini, G., Venier, I., Parolaro, D., Zuccaro, P., and Gori, E. (1994). Effect of morphine and methadone acute treatment on immunological activity in mice: Pharmacokinetic and pharmacodynamic correlates. *J. Pharmacol. Exp. Ther.* **269**:1112–1116.

44. Bessler, H., Sztein, M. B., and Serrate, S. A. (1990). Beta-endorphin modulation of IL-1-induced IL-2 production. *Immunopharmacology* **19**:5–14.

45. van den Bergh, P., Dobber, R., Ramlal, S., Rozing, J., and Nagelkerken, L. (1994). Role of opioid peptides in the regulation of cytokine production by murine CD4+ T cells. *Cell. Immunol.* **154**:109–122.

46. Ryng, S., Zimecki, M., Sonnenberg, Z., and Mokrosz, M. J. (1999). Immunomodulating action and structure–activity relationships of substituted phenylamides of 5-amino-3-methylisoxazole-4-carboxylic acid. *Arch. Pharm. (Weinheim)* **332**:158–162.

47. Colasanti, M. and Suzuki, H. (2000). The dual personality of NO. *Trends Pharmacol. Sci.* **21**:249–252.

48. MacMicking, J., Xie, Q. W., and Nathan, C. (1997). Nitric oxide and macrophage function. *Ann. Rev. Immunol.* **15**:323–350.

49. Haynes, B. F. and Hale, L. P. (1999). Thymic function, aging, and AIDS. *Hosp. Pract. (Off Ed).* **34**:59–60, **63–5**:69–70.

50. Wang, C., McFadyen, I. J., Traynor, J. R., and Mosberg, H. I. (1998). Design of a high affinity peptidomimetic opioid agonist from peptide pharmacophore models. *Bioorg. Med. Chem. Lett.* **8**:2685–2688.

51. Roda, L. G., Bongiorno, L., Trani, E., Urbani, A., and Marini, M. (1996). Positive and negative immunomodulation by opioid peptides. *Int. J. Immunopharmacol.* **18:**1–16.

52. Lysle, D. T., Coussons, M. E., Watts, V. J., Bennett, E. H., and Dykstra, L. A. (1993). Morphine-induced alterations of immune status: Dose dependency, compartment specificity and antagonism by naltrexone. *J. Pharmacol. Exp. Ther.* **265:**1071–1078.

53. Muñoz-Fernandez, M. A. and Fresno, M. (1998). The role of tumour necrosis factor, interleukin 6, interferon-gamma and inducible nitric oxide synthase in the development and pathology of the nervous system. *Prog. Neurobiol.* **56:**307–340.

54. Albina, J. E., Cui, S., Mateo, R. B., and Reichner, J. S., (1993). Nitric oxide-mediated apoptosis in murine peritoneal macrophages. *J. Immunol.* **1150:**5080–5085.

55. Stein, C. S. and Strejan, G. H. (1993). Rat splenocytes inhibit antigen-specific lymphocyte proliferation through a reactive nitrogen intermediate (RNI)-dependent mechanism and exhibit increased RNI production in response to IFN-gamma. *Cell. Immunol.* **150:**281–297.

56. Raber, J. and Bloom, F. E. (1994). IL-2 induces vasopressin release from the hypothalamus and the amygdala: Role of nitric oxide-mediated signaling. *J. Neurosci.* **14:**6187–6195.

57. Dunn, A. J. and Wang, J. (1995). Cytokine effects on CNS biogenic amines. *Neuroimmunomodulation* **2:**319–328.

58. Clarke, C. J., Hales, A., Hunt, A., and Foxwell, B. M. (1998). IL-10-mediated suppression of TNF-alpha production is independent of its ability to inhibit NF kappa B activity. *Eur. J. Immunol.* **28:**1719–1726.

59. Hambrook, J. M., Morgan, B. A., Rance, M. J., and Smith, C. F. (1976). Mode of deactivation of the enkephalins by rat and human plasma and rat brain homogenates. *Nature* **262:**782–783.

60. Carr, D. J., Rogers, T. J., and Weber, R. J. (1996). The relevance of opioids and opioid receptors on immunocompetence and immune homeostasis. *Proc. Soc. Exp. Biol. Med.* **213:**248–257.

13

Effect of Alcohol on Microbial Infection

YOSHIMASA YAMAMOTO and HERMAN FRIEDMAN

1. INTRODUCTION

It is widely accepted that alcohol is one of the leading health problems in the United States. It has been pointed out that there are at least 8–10 million alcoholics in the United States and more than one-half million deaths per year are attributed to alcohol abuse and alcohol-associated illnesses. Many of these deaths are associated with infections, as well as with neoplasm and liver injury. Impairment by alcohol of host defenses to infections has been recorded since the late 18th century when "ardent spirits" were noted to increase consumption, that is, tuberculosis, as well as pneumonia and yellow fever. Koch, in 1884, reported that intoxicated rats were more susceptible to cholera infection than normal rats.[1] Experimental evidence continues to accumulate since those early studies and several definitive reports have provided confirmation of a connection between alcohol consumption of humans and pneumonia. Furthermore, the Addiction Research Foundation in Toronto has provided population-based rates of pneumonia mortality in alcoholics vs normal subjects.[2] They reported that the ratio observed of expected deaths due to infection for alcoholic men was three times greater than for controls, and for alcoholic women seven times greater. Another study based on nine hundred consecutive admissions to Yale-New Haven Hospital showed that 16% of alcoholics vs 6.5% of nonalcoholics have bacterial pneumonia.[3] Recent study also indicates that heavy alcohol

YOSHIMASA YAMAMOTO • Department of Basic Laboratory Sciences, Osaka University Graduate School of Medicine, Osaka, Japan. HERMAN FRIEDMAN • Department of Medical Microbiology and Immunology, University of South Florida College of Medicine, Tampa, FL 33612.

Infectious Diseases and Substance Abuse, edited by Herman Friedman *et al.*
Springer, New York, 2005.

consumption increases the risk of nosocomial infection in men who underwent general surgical procedures.[4]

Alcohol, as well as other drugs of abuse, has been suggested as a possible cofactor in the development of AIDS by several investigators.[5] This speculation is based upon reports of immune suppression by alcohol, as well as development of life-threatening infections associated with immunocompromised hosts. Thus, there is a strong basis for suspecting alcohol consumption may have a negative impact on the immune response in man in clinical situations. Ethanol-induced suppression of the immune system has been studied in detail in animal model systems.[6–8] For example, Jerrells *et al.* reported that rats administered ethanol show depressed immune functions.[7] In other studies, alterations in immunity were correlated with changes in steroid levels which occurred during the withdrawal phase of intoxication.[9] Thus, both a direct and indirect effect of ethanol on immune functions have been recorded.

Bermudez and Young reported that infections caused by organisms belonging to the *Mycobacterium avium* complex are associated with monocytes or macrophage dysfunction induced by ethanol.[10] Cultured human monocyte-derived macrophages and murine Kupffer cells exposed to appropriate concentrations of ethanol showed greater intracellular growth of *M. avium* than did control cells. Furthermore, lymphokines such as tumor necrosis factor (TNF) or granulocyte-macrophage colony-stimulating factor (GM-CSF) were less effective in inducing killing activity of the macrophages against *Mycobacterium* in the presence of ethanol. Further, mice given ethanol and infected intravenously with *Mycobacterium* showed greater numbers of these bacteria in their blood, liver, and spleen than controls and progressive infection. Saad *et al.*[11] also showed that mice fed ethanol by the Lieber–DiCarli diet became susceptible to infection with *Listeria monocytogenes*. There was a greater accumulation of these intracellular bacterial pathogens in the liver of alcohol-fed animals, as shown by the higher number of bacteria recovered from the liver as well as spleen of the animals given alcohol and challenged with these bacteria. Similar findings in rodent model of BCG infection were also reported.[12] Thus, a possible alcohol effect on the resistance to bacterial infection appeared likely.

2. ALCOHOL ON *LEGIONELLA PNEUMOPHILA* INFECTION

L. pneumophila, an intracellular opportunistic gram-negative pathogen, which infects primarily macrophages and is an etiologic cause of serious pneumonia in immunocompromised individuals, has been utilized for determination of effect of alcohol on bacterial infections. An animal infection model has been developed which permitted study of the mechanism of resistance to this organism. The inbred mouse strain A/J was shown to be more susceptible to *L. pneumophila* than other strains of mice, such as BALB/c mice.[13] The A/J mouse model was found to be extremely useful to study infection by this organism since the immune response of these mice to *L. pneumophila* involved macrophage responsiveness similar to that of human.[14] Furthermore, the interaction between macrophages and *L. pneumophila* is controlled by cytokines.[15] Thus, the regulation of both macrophages and cytokine production was found critical for host resistance to *L. pneumophila* infection.

Treatment of A/J mice, as well as resistant BALB/c mice, *in vivo* with doses of ethanol similar to the range found in humans who are alcoholics was shown to affect their resistance to this organism (unpublished data). The *L. pneumophila* infection susceptible A/J mice and resistant BALB/c mice fed the Liber–DiCarli liquid diet containing ethanol (35% ethanol derived calories) were shown to have essentially a similar consumption of the diet. For example, control mice given the same volume of non-ethanol containing diet as consumed by the experimental mice, showed that the nutritional conditions between ethanol-treated vs control mice were similar. Therefore, at least in terms of the level of food intake, the A/J mice consumed the same amount of ethanol as BALB/c mice.

Treating the A/J mice with ethanol for up to 7 days only slightly increased their genetically determined susceptibility to infection. However, feeding BALB/c mice with ethanol-containing diet significantly increased the susceptibility of these mice to *L. pneumophila* infection. For example, 7 days of feeding the animals with ethanol resulted in an obvious increase in susceptibility of the mice to pulmonary infection with *L. pneumophila*. Furthermore, the number of leukocytes in alveolar lavage fluid obtained from the mice inoculated with *L. pneumophila* intratracheally and administered ethanol orally showed restricted immigration of neutrophils. However, the mechanism of altered susceptibility of the mice to *L. pneumophila* infection induced by ethanol feeding is not clear and further studies should investigate the mechanisms involved. It seems likely that study of the genetically susceptible A/J mice in comparison to resistant BALB/c mice will provide useful information concerning the mechanisms involved in the effect of alcohol on altered susceptibility to an opportunistic infectious agent such as *L. pneumophila*, which is associated with pulmonary disease and pneumonia, especially in immunocompromised individuals.

3. ALCOHOL ON *IN VITRO* SUSCEPTIBILITY OF MACROPHAGES TO *L. PNEUMOPHILA* INFECTION

The addition of various doses of ethanol to cultures from *L. pneumophila* infected macrophages resulted in divergent effects, based on genotype of the mouse strain examined in terms of susceptibility or resistance to *L. pneumophila*.[16] The addition of ethanol to macrophage cultures from susceptible A/J mice significantly suppressed the ability of the cells to replicate bacteria as normally occurs in macrophages from untreated A/J mice. For example, normal cultures without ethanol treatment evinced a 100-fold increase in the number of viable bacteria determined by colony forming units (CFU) assay within 2 days when the macrophages are infected with bacteria. An infectivity ratio of 10 bacteria per 1 macrophage was used, which has been known to be optimal for *in vitro* infection. In the presence of 0.5% v/v (85 mM) ethanol, it was found that there was a significant suppression in the ability of bacteria to replicate in these macrophages when they were pretreated with ethanol for 3 hr before infection (47 ± 2% suppression, $p < 0.005$). However, a higher concentration of ethanol (1.0% v/v, 170 mM) resulted in relatively less activity regarding growth of bacteria, as compared to lower concentrations of ethanol, that is, a biphasic ethanol dose response curve was apparent.

The reason for a biphasic ethanol dose response curve, however, was not clear. It is possible that the higher concentration of ethanol could increase toxicity to the macrophages, although there was no evidence this occurred, at least in terms of morphology of the cells or trypan blue stain for viability. The uptake of *L. pneumophila* by the ethanol-treated macrophages was found to be essentially similar to that of untreated macrophages in terms of viability count at the time of *in vitro* infection. Furthermore, 24-hr pretreatment with ethanol, a relatively long incubation period with ethanol compared with 3-hr pretreatment as mentioned above, did not induce a significant change in the number of viable bacteria in the macrophages 24 hr after infection. Nevertheless, it is apparent that ethanol treatment of the macrophages can alter the innate ability of the cells to be infected with *L. pneumophila*.

The results obtained with resistant BALB/c mice, which do not replicate *L. pneumophila* well *in vitro*, contrasted markedly with results of similar studies with macrophages from susceptible A/J mice. For example, there was only a maximum 2–5-fold increase in *L. pneumophila* colonies within 2 days after infection of similar numbers of macrophages from BALB/c mice with the same number of bacteria as compared to the 100-fold or greater increase of bacteria in macrophages from the A/J mice. This difference was not related to the presence or absence of virulence factors in the *L. pneumophila*, since the same organisms and the same culture conditions utilized with macrophages from both strains of mice resulted in these differences. The only obvious difference was the sources of macrophages, that is, a *L. pneumophila* resistant mouse strain vs a susceptible strain. When ethanol was used at the same concentration used with the macrophage from A/J mice, there was an increase rather than a decrease in growth of bacteria in the cultures of macrophages from the BALB/c mice treated with 0.1–0.5% (v/v) concentrations of ethanol. These concentrations of ethanol, which are significantly below the toxic levels to macrophages, increased rather than decreased the ability of the cells to replicate bacteria. Such observations indicate that ethanol could suppress resistance mechanisms of the macrophages to *L. pneumophila* growth otherwise *in vitro* in nonpermissive macrophages and also suppress the ability of the bacteria to grow in normally susceptible mouse macrophages. These differences may reflect differences in the nature of the macrophages from permissive vs nonpermissive individuals in terms of replication of opportunistic microorganisms such as *L. pneumophila*, which grows preferentially in macrophages.

Since macrophages from BALB/c mice, as well as from other nonpermissive mice, do not permit significant replication of bacteria,[13] this may be due to an innate inability of the phagocytic cells from these strains of mice to provide an appropriate environment for growth of the bacteria because of nutritional mechanisms or differences. Thus, macrophages from these nonsusceptible mouse strains evince a genetic predisposition to mount an inhibitory effect against *L. pneumophila* growth.[17,18] It is therefore possible that metabolic pathways in macrophages directly related to antimicrobial activities could be affected by alcohol treatment and thus result in inhibition of macrophage activity, permitting greater bacterial replication. This is similar to the observation by Bermudez and Young[10] that ethanol augments the intracellular survival of *M. avium* complex in macrophage cultures. It seems likely from the similar studies with *L. pneumophila* that ethanol treatment of macrophage cultures significantly alters the ability of the cells to kill or inhibit growth of the microbes *in vitro*. Concentrations of ethanol required for induction of macrophage dysfunction, however, are relatively high compared with the concentrations of ethanol used for studies *in vivo*. It seems clear that ethanol can induce a dysfunction of macrophages in terms of growth of opportunistic bacteria like *L. pneumophila* in these cells.

4. ALCOHOL ON *IN VITRO* CYTOKINE RESPONSE OF MACROPHAGES TO MICROBES

It has been known that prolonged and excessive consumption of alcohol results in alterations of host immunity.[19] For example, impaired immunity in alcoholics was reported with regard to both humoral immune mechanisms, including antibody production,[20] and various aspects of cell-mediated immunity, such as delayed-type hypersensitivity reactions[21,22] and lymphocyte proliferative responses to mitogens.[8,23] Furthermore, studies with experimental animals were shown that administration of alcohol markedly suppresses many immune functions.[7,24–27] Such experimental studies suggest deleterious effects of alcohol on the immune system. However, the direct effects of alcohol on immune cells is not well understood due to the complexity of the immune response system.

Determination of possible direct effects of alcohol on immune cells is essential for understanding the mechanism of immunodisturbance and higher incidence of infections in alcoholics. Since macrophages play a pivotal role in the generation of immunity to invading microorganisms, as well as in inflammatory responses in general, the possible effect of alcohol on macrophages is critical to an understanding of the mechanism of increased susceptibility of alcoholics to infection. In this regard, there are many reports regarding alcohol effects *in vitro* on monocytes/macrophages. For example, the *in vitro* exposure of human monocytes to a single, short-term alcohol treatment results in decreased TNF production capacity.[28–30] IL-1β and IL-6 production of monocytes, both at the level of mRNA and protein secretion, are affected by ethanol treatment *in vitro*.[31,32] The altered production of GM-CSF in human monocytes by ethanol has also been reported.[28] Thus, it seems likely that alcohol can directly affect cytokine production by monocytes/macrophages *in vitro*. However, the details of alcohol effects on macrophages regarding immunomodulatory activity are still unknown, since pathogen–macrophage interactions are complicated and not well understood. In this regard, our study concerning *Candida albicans*–macrophage interaction in terms of cytokine and chemokine induction provides a useful model for study of alcohol effects on immune cells.

4.1. A Model for Analysis of Cytokine and Chemokine Induction of Macrophages

C. albicans, a pathogenic yeast which causes candidiasis, including pulmonary infection, has a relatively simple outer structure compared with other microorganisms such as gram-negative bacteria. Mannan is a major component of Candida cell wall and, as shown by our study,[33] causes cytokine induction by Candida attachment to macrophages, which might be a useful model for analysis of alcohol effects on immune cells.

When mouse macrophages were stimulated with *C. albicans* in the presence of cytochalasin D, which prevents uptake of microorganisms by macrophages but permits the attachment of the microbes to the macrophages,[34] increased steady-state levels of cytokine (IL-1β, IL-6, and GM-CSF) and chemokine (MIP-1β, MIP-2, and KC) mRNAs were measured by quantitative RT-PCR (Fig. 1). These data indicate that Candida attachment to macrophages is sufficient to generate signals for increasing cytokine and chemokine messages, as observed for other pathogens.[34]

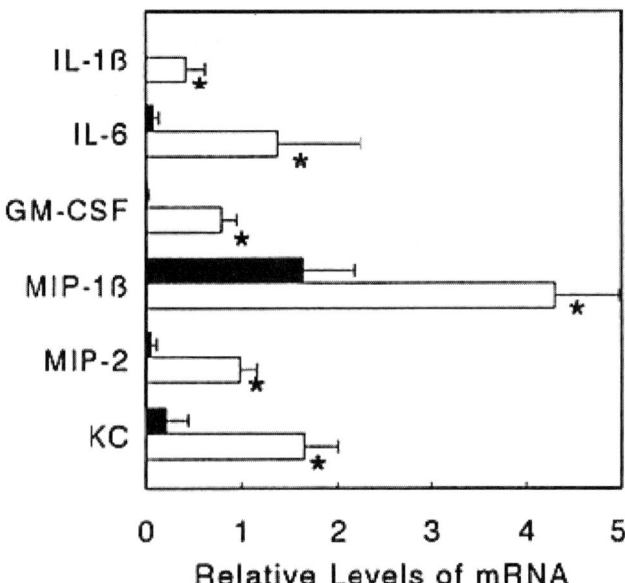

FIGURE 1. Levels of cytokine and chemokine mRNAs in macrophages incubated with or without *C. albicans* as determined by quantitative RT-PCR. Macrophages were pretreated with cytochalasin D for 30 min and then incubated with or without *C. albicans* for 1 hr in the presence of cytochalasin D. The ratio of yeast cells to macrophages was 1 : 10. Each bar represents the mean (error bar indicates ± standard error) for three experiments. ■, Control; □, with *C. albicans*; *, $P < 0.05$ compared with control group. From ref. [33].

4.2. Different Receptor/Signaling Pathways Involved in Chemokine/Cytokine Induction by Candida Attachment

The analysis of the receptor/signaling pathways involved in cytokine and chemokine induction by Candida attachment on mRNA induction utilizing 3α-methyl-D-mannoside (αMM), which has a high affinity for carbohydrate–protein interaction, especially in mannose related interactions, as well as protein kinase inhibitors, showed that cytokine- and chemokine-inducing systems, including ligand/receptor interactions, by Candida attachment are different. That is, αMM treatment markedly reduced the induction of cytokine (GM-CSF) mRNA by Candida attachment, but induction of chemokine (MIP-2) mRNA was not affected. Studies using protein kinase inhibitors showed that the GM-CSF inducing pathway is calmodulin and myosin light chain kinase dependent, but the MIP-2 pathway is not (unpublished data).

4.3. Receptors in Cytokine and Chemokine Induction

Since αMM has a strong inhibitory effect on cytokine mRNA induction, the possible involvement of the mannose receptor in cytokine and chemokine induction was examined using the antisense oligonucleotide technique. Treatment of macrophages with the antisense phosphorothioate oligodeoxyribonucleotides (20 bp) of mannose receptor, which hybridized to the 3′-untranslated region, 4801–4820 of mannose-receptor mRNA, showed the best result among six different antisense oligonucleotides tested, including the initiation of translation region. There was a significant decrease of mannose-receptor expression and functional activity measured by western blotting and ^{125}I-labeled mannose-BSA uptake, respectively. When mannose receptor downregulated macrophages were stimulated with Candida in the presence of cytochalasin D, induction of cytokine mRNAs, such as IL-1β, IL-6, and GM-CSF, was markedly reduced compared with normal macrophages stimulated with Candida. In contrast, chemokine mRNAs, such as MIP-1α, MIP-2, and KC, were not affected by the mannose-receptor downregulation. These data clearly showed that cytokine induction by Candida attachment is mediated by mannose receptor, but chemokine induction is not.[33] On the other hand, inhibition of chemokine induction, not cytokine induction, with scavenger receptor inhibitors, such as dextran sulfate, fucoidan, and poly (I), indicates a possible involvement of scavenger receptor in chemokine induction by Candida attachment.

4.4. Effect of Ethanol on Cytokine and Chemokine Induction

Treatment of macrophages with 100 mM ethanol (1 hr pretreatment and 1 hr post-Candida stimulation) showed marked inhibition of GM-CSF and IL-6 mRNA induction by Candida attachment, but inhibition of MIP-1β and MIP-2 mRNA induction was minimum (Fig. 2). This selective inhibition of cytokine mRNA induction was also observed when L. pneumophila was used as an infectant. A dose response study of ethanol (25–500 mM) on GM-CSF mRNA

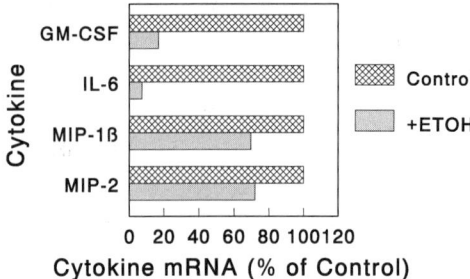

FIGURE 2. Cytokine and chemokine induction by Candida attachment to macrophages. Thioglycolate-elicited mouse (BALB/c) peritoneal macrophages were pretreated with 1 μg/ml of cytochalasin D for 30 min and then stimulated with Candida in the presence of cytochalasin D for 1 hr. The ratio of yeast cells to macrophages was 1 : 10. Total RNA was isolated and subjected to quantitative RT-PCR. Data normalized relative to an endogenous standard (β2-microglobulin, BMG) by comparing the ratios of PCR products and expressed as cytokine or chemokine levels/BMG level.

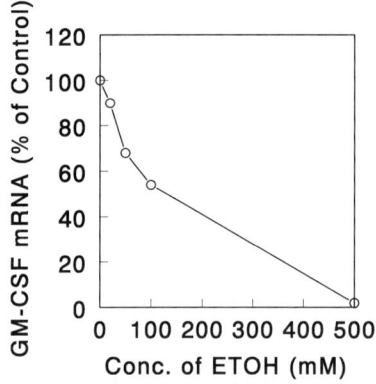

FIGURE 3. Effect of ethanol treatment on GM-CSF mRNA levels induced by *L. pneumophila* attachment to macrophages. Macrophages were treated with various concentrations of ethanol for 1 hr and then stimulated with bacteria in the presence of cytochalasin D and ethanol for 1 hr. Total RNA was isolated and subjected to quantitative RT-PCR. Data normalized relative to an endogenous standard (β2-microglobulin, BMG) by comparing the ratios of PCR products and expressed as cytokine or chemokine levels/BMG level.

induction by *L. pneumophila* attachment to macrophages showed that even 50 mM ethanol inhibited the induction of GM-CSF mRNA (Fig. 3).

The results obtained by studies mentioned above showed that (1) Candida attachment to macrophages induces increased levels of cytokine and chemokine mRNA, which are mediated by both a mannose-receptor/signaling pathway for cytokine and a scavenger receptor/signaling pathway for chemokine, respectively. (2) Alcohol treatment of macrophages induces a selective inhibition of cytokine mRNA increase by both Candida and Legionella attachment. That is, target site(s) of alcohol in macrophages may be common in cytokine induction between Candida and Legionella attachment, such as activation of a transcription factor and/or other factors involved in signal transduction. Cytokine induction by Candida attachment is mediated by mannose receptors, but Legionella attachment may be mediated by a different receptor, because our study showed that flagella (flagellin protein polymer) is one of the major bacterial ligands for induction of cytokine, but not chemokine, by gram-negative bacteria attaching to macrophages.[35] Since it has been demonstrated that induction of cytokine messages, including IL-1β, IL-6, and GM-CSF, is regulated by a common transcription factor, NFκB, it seems likely that such a common nuclear transactivating factor or other common pathway between different receptor-mediated cytokine induction, which is not shared with chemokine induction, may be a target site of alcohol.

5. CONCLUSION

The studies to date suggest that alcohol is one of the many agents with low or negligible toxicity which can alter the ability of a host to resist an opportunistic infection caused by an organism, such as *L. pneumophila*. In all probability, an individual with a fully competent immune response system, and evincing normal resistance mechanisms would probably not be susceptible to infection with a particular pathogen, such as *L. pneumophila*. However, an individual who is already immunocompromised due to concomitant or preexisting infection with an immunosuppressive virus such as HIV or other microbes may be more susceptible to an opportunistic pathogen when exposed to a drug of abuse, including alcohol, which has the potential of dysregulating immunity. Obviously more studies should be performed since there are now many tools available to examine the nature and mechanism of resistance to microorganisms at the subcellular and genetic level.

REFERENCES

1. Smith, F. E. and Palmer, D. L. (1976). Alcoholism, infection and altered host defenses: A review of clinical and experimental observations. *J. Chronic Dis.* **29**:35–49.
2. Schmidt, W. and De Lint, J. (1972). Causes of death of alcoholics. *Q. J. Stud. Alcohol* **33**:171–185.
3. Nolan, J. P. (1965). Alcohol as a factor in the illness of university service patients. *Am. J. Med. Sci.* **249**:135–142.
4. Delgado-Rodriguez, M., Mariscal-Ortiz, M., Gomez-Ortega, A., Martinez-Gallego, G., Palma-Perez, S., Sillero-Arenas, M. *et al.* (2003). Alcohol consumption and the risk of nosocomial infection in general surgery. *Br. J. Surg.* **90**:1287–1293.
5. Siegel, L. (1986). AIDS: Relationship to alcohol and other drugs. *J. Subst. Abuse Treat.* **3**:271–274.
6. Caiazza, S. S. and Ovary, Z. (1976). Effects of ethanol intake on the immune system of guinea pigs. *J. Stud. Alcohol.* **37**:959–964.
7. Jerrells, T. R., Marietta, C. A., Eckardt, M. J., Majchrowicz, E., and Weight, F. F. (1986). Effects of ethanol administration on parameters of immunocompetency in rats. *J. Leukoc. Biol.* **39**:499–510.

8.   Roselle, G. A. and Mendenhall, C. L. (1984). Ethanol-induced alterations in lymphocyte function in the guinea pig. *Alcohol Clin. Exp. Res.* **8**:62–67.

9.   Tabakoff, B., Jafee, R. C., and Ritzmann, R. F. (1978). Corticosterone concentrations in mice during ethanol drinking and withdrawal. *J. Pharm. Pharmacol.* **30**:371–374.

10. Bermudez, L. E. and Young, L. S. (1991). Ethanol augments intracellular survival of Mycobacterium avium complex and impairs macrophage responses to cytokines. *J. Infect. Dis.* **163**:1286–1292.

11. Saad, A. J., Domiati-Saad, R., and Jerrells, T. R. (1993). Ethanol ingestion increases susceptibility of mice to Listeria monocytogenes. *Alcohol Clin. Exp. Res.* **17**:75–85.

12. Mendenhall, C. L., Grossman, C. J., Roselle, G. A., Ghosn, S., Gartside, P. S., Rouster, S. D. *et al.* (1990). Host response to mycobacterial infection in the alcoholic rat. *Gastroenterology* **99**:1723–1726.

13. Yamamoto, Y., Klein, T. W., Newton, C. A., Widen, R., and Friedman, H. (1988). Growth of Legionella pneumophila in thioglycolate-elicited peritoneal macrophages from A/J mice. *Infect. Immun.* **56**:370–375.

14. Yamamoto, Y., Klein, T. W., Newton, C., and Friedman, H. (1992). Differing macrophage and lymphocyte roles in resistance to Legionella pneumophila infection. *J. Immunol.* **148**:584–589.

15. Klein, T. W., Yamamoto, Y., Brown, H. K., and Friedman, H. (1991). Interferon-gamma induced resistance to Legionella pneumophila in susceptible A/J mouse macrophages. *J. Leukoc. Biol.* **49**:98–103.

16. Yamamoto, Y., Klein, T. W., and Friedman, H. (1993). Differential effects of ethanol on permissive versus nonpermissive macrophages infected with *Legionella pneumophila*. *Proc. Soc. Exp. Biol. Med.* **203**:323–327.

17. Yamamoto, Y., Klein, T. W., and Friedman, H. (1992). Genetic control of macrophage susceptibility to infection by Legionella pneumophila. *FEMS Microbiol. Immunol.* **4**:137–145.

18. Yamamoto, Y., Klein, T. W., and Friedman, H. (1991). *Legionella pneumophila* growth in macrophages from susceptible mice is genetically controlled. *Proc. Soc. Exp. Biol. Med.* **196**:405–409.

19. MacGregor, R. R. (1986). Alcohol and immune defense. *JAMA* **256**:1474–1479.

20. Smith, W. I., Jr., Van Thiel, D. H., Whiteside, T., Janoson, B., Magovern, J., Puet, T. *et al.* (1980). Altered immunity in male patients with alcoholic liver disease: Evidence for defective immune regulation. *Alcohol Clin. Exp. Res.* **4**:199–206.

21. Berenyi, M. R., Straus, B., and Cruz, D. (1974). In vitro and in vivo studies of cellular immunity in alcoholic cirrhosis. *Am. J. Dig. Dis.* **19**:199–205.

22. Gluckman, S. J., Dvorak, V. C., and MacGregor, R. R. (1977). Host defenses during prolonged alcohol consumption in a controlled environment. *Arch. Intern. Med.* **137**:1539–1543.

23. Young, G. P., Van der Weyden, M. B., Rose, I. S., and Dudley, F. J. (1979). Lymphopenia and lymphocyte transformation in alcoholics. *Experientia* **35**:268–269.

24. Abdallah, R. M., Starkey, J. R., and Meadows, G. G. (1983). Alcohol and related dietary effects on mouse natural killer-cell activity. *Immunology* **50**:131–137.

25. Jayasinghe, R., Gianutsos, G., and Hubbard, A. K. (1992). Ethanol-induced suppression of cell-mediated immunity in the mouse. *Alcohol Clin. Exp. Res.* **16**:331–335.

26. Mendenhall, C. L., Finkelman, F., Means, R. T., Jr., Sherman, K. E., Nguyen, V. T., Grossman, C. E. *et al.* (1999). Cytokine response to BCG infection in alcohol-fed mice. *Alcohol* **19**:57–63.

27. Sibley, D. A., Osna, N., Kusynski, C., Wilkie, L., and Jerrells, T. R. (2001). Alcohol consumption is associated with alterations in macrophage responses to interferon-gamma and infection by *Salmonella typhimurium*. *FEMS Immunol. Med. Microbiol.* **32**:73–83.

28. Bermudez, L. E., Wu, M., Martinelli, J., and Young, L. S. (1991). Ethanol affects release of TNF and GM-CSF and membrane expression of TNF receptors by human macrophages. *Lymphokine Cytokine Res.* **10**:413–419.

29. Strieter, R. M., Remick, D. G., Ham, J. M., Colletti, L. M., Lynch, J. P., 3rd, and Kunkel, S. L. (1990). Tumor necrosis factor-alpha gene expression in human whole blood. *J. Leukoc. Biol.* **47**:366–370.

30. Verma, B. K., Fogarasi, M., and Szabo, G. (1993). Down-regulation of tumor necrosis factor alpha activity by acute ethanol treatment in human peripheral blood monocytes. *J. Clin. Immunol.* **13**:8–22.

31. Szabo, G., Verma, B., and Catalano, D. (1993). Selective inhibition of antigen-specific T lymphocyte proliferation by acute ethanol exposure: The role of impaired monocyte antigen presentation capacity and mediator production. *J. Leukoc. Biol.* **54**:534–544.

32. Yamamoto, Y., Klein, T. W., and Friedman, H. (1993). Ethanol affects macrophage production of IL-6 and susceptibility to infection by *Legionella pneumophila. Adv. Exp. Med. Biol.* **335**:169–173.

33. Yamamoto, Y., Klein, T. W., and Friedman, H. (1997). Involvement of mannose receptor in cytokine interleukin-1beta (IL-1beta), IL-6, and granulocyte-macrophage colony-stimulating factor responses, but not in chemokine macrophage inflammatory protein 1beta (MIP-1beta), MIP-2, and KC responses, caused by attachment of *Candida albicans* to macrophages. *Infect. Immun.* **65**:1077–1082.

34. Yamamoto, Y., Okubo, S., Klein, T. W., Onozaki, K., Saito, T., and Friedman., H. (1994). Binding of *Legionella pneumophila* to macrophages increases cellular cytokine mRNA. *Infect. Immun.* **62**:3947–3956.

35. Yamamoto, Y., Klein, T. W., and Friedman, H. (1996). Induction of cytokine granulocyte-macrophage colony-stimulating factor and chemokine macrophage inflammatory protein 2 mRNAs in macrophages by *Legionella pneumophila* or *Salmonella typhimurium* attachment requires different ligand-receptor systems. *Infect. Immun.* **64**:3062–3068.

14

Brucella Infection and Ethanol

ZEKI YUMUK

Ethanol exposure adversely affects the infection of rats caused by *Brucella melitensis*[1] and brucellosis therapy may be ineffective (unpublished result). An understanding of the interaction between ethanol and *Brucella* infection is crucial. Not only will it augment our knowledge of how the brucellosis is affected by the chronic ethanol consumption, but it may also lead to new sights on the therapy of brucellosis.

1. *BRUCELLA* INFECTION

Brucellosis is a zoonotic infection of domesticated and wild animals, caused by organisms of the genus *Brucella*. Human infection by *Brucella* spp. still constitutes an important health problem in many countries and in some developed areas of the world. The organism infects mainly cattle, sheep, goats, and other ruminants in which it causes abortion, fatal death, and genital infection. Humans, who are infected incidentally by contact with infected animals or ingestion of dairy foods, may develop numerous symptoms in addition to the usual ones of fever, malaise, and muscle pain. Disease frequently becomes chronic and may relapse, even with treatment. This infection is considered to be a problem, because *Brucella abortus* vaccines do not protect effectively against *B. melitensis* infection.[2] Moreover, the ease of transmission by aerosol suggests that *Brucella* organisms might be a potential candidate for use as a biological warfare agent.[3]

ZEKI YUMUK • Department of Clinical Microbiology, Kocaeli University Faculty of Medicine, Kocaeli, Turkey.

Infectious Diseases and Substance Abuse, edited by Herman Friedman *et al.*
Springer, New York, 2005.

The main pathogenic species worldwide are *B. melitensis* and *B. abortus*. The true incidence of human brucellosis is unknown. Reported incidence in endemic disease areas varies widely, from <0.01 to >200 per 100,000 population.[2] The disease exists worldwide, especially in the Mediterranean basin, the Arabian Peninsula, the Indian subcontinents, part of Mexico, and Central and South America. Prevention of human brucellosis depends on the control of the disease in animals. The greatest success has been achieved in eradicating the bovine disease, mainly in industrialized countries. *B. melitensis* infection has proven more intractable, and success has been limited.[2]

Brucellae are small, nonmotile, nonsporulating, nontoxigenic, nonfermenting, aerobic, Gram-negative coccobacilli that may, based on DNA homology, represent a single species.[4,5] However, they are classified into seven species according to antigenic variation and primary host: *B. melitensis* (sheep and goats), *Brucella suis* (hogs), *B. abortus* (cattle), *Brucella ovis* (sheep), *Brucella canis* (dogs), *Brucella neotomae* (wood rats), and *Brucella maris* (marine mammals). *B. abortus* and *B. canis* tend to produce mild diseases with rare suppurative complications. *B. melitensis*, the most common cause of brucellosis, also causes severe diseases with a high incidence of serious complications.

Brucella can enter a mammalian host through skin abrasions or cuts, the conjunctiva, the respiratory tract, and the gastrointestinal tract. In the gastrointestinal tract, the organisms are phagocytosed by M cells, from which they gain access to the submucosa. Organisms are rapidly ingested by polymorphonuclear leukocytes, which generally fail to kill them, and are also phagocytosed by macrophages, which traffic to lymphoid tissue draining the infection site, and may eventually localize in lymph nodes, liver, spleen, mammary glands, joints, kidneys, and bone marrow. Survival in macrophages, considered to be responsible for the establishment of chronic infections, allows the bacteria to escape the extracellular mechanisms of host defense, such as complement and antibodies.

The clinical picture in human brucellosis can be misleading. The spectrum of human brucellosis ranges from subclinical to chronic. Symptoms are non-specific, usually occurring within 2–3 weeks of inoculation. The onset of disease is insidious in approximately one half of cases. Chronically infected patients frequently lose weight. Symptoms often last for 3–6 months and occasionally for a year or more. Physical examination is usually normal, despite the occurrence of hepatomegaly, splenomegaly, or lymphadenopathy. Poor diagnosis and treatment may result in serious, sometimes life-threatening complications such as spondylitis, infectious endocarditis, and encephalitis.

When the disease is considered, diagnosis is usually made by serology. Although a number of serological techniques have been developed and tested, the tube agglutination test still remains the standard method. The tube agglutination test does not detect antibodies to *B. canis*. In addition to serologic testing, diagnosis should be pursued by microbiologic culture of blood or body fluid samples. Because it is extremely infectious for laboratory workers, the organism should be subcultured only in a biohazard hood.

Therapy with single drug has resulted in a high relapse rate, so combined regimens should be used. Antimicrobial therapy of brucellosis relieves symptoms, shortens the duration of illness, and reduces the incidence of complications, some of which can be life threatening. The treatment recommended by the World Health Organization for acute brucellosis in adults is rifampicin 600–900 mg and doxycycline 200 mg daily for a minimum of 6 weeks. Some still claim that the long-established combination of intramuscular streptomycin with an oral tetracycline gives fewer relapses.[6] *In vitro* antimicrobial susceptibility tests reveal that a variety of agents have activity against *Brucella*.[7] Since brucellosis is an intracellular infection, the use of *in vitro* susceptibility tests for *B. melitensis* may not accurately predict the therapeutic results in human infections.[8]

2.   THE EFFECT OF LONG-TERM ETHANOL FEEDING

Infections, neoplasm, and chronic liver injury are common causes of morbidity and mortality in alcoholics, and all these may be related to an underlying altered immune response.[9] With respect to infections, it is generally thought that immunodeficiency caused by alcohol abuse is the major factor. Alteration in innate immunity and adaptive immune responses after alcohol use can lead to decreased host defenses and increased susceptibility to infections.[10] Once such infections develop, they are usually more severe, and some are associated with a higher mortality than that found in the nonalcoholic population.[11]

Animal models of infections in the presence or absence of ethanol exposure have demonstrated a number of important findings. Ethanol exposure adversely affects the infections of experimental animals caused by bacteria such as *Listeria monocytogenes*,[12] *Streptococcus pneumonia*,[13] *Legionella pneumophila*,[14] *Mycobacterium avium* complex.[15]

The observations using the model described previously [1] are consistent with an alcohol-induced increase in host susceptibility to *Brucella* infection. It was found that the chronically ethanol-receiving rats exposed to *B. melitensis* infection had significantly ($p < 0.01$) greater number of *B. melitensis* in their spleen and liver than the rats in the control group (Fig. 1). However, although there was a moderate relationship between the amount of ethanol consumption and the number of *B. melitensis* in spleen ($r = -0.062$, $p > 0.05$), it was considered insignificant (Fig. 2). There were no physical signs of infection observed in rats after they were challenged by *B. melitensis*. In order to show any possible enlargement of spleen and liver, the organs to body weights ratio was calculated. There were no significant differences found for the spleen ($p = 0.204$) and liver ($p = 0.977$)—body ratio between groups.

Host resistance to intracellular parasites is associated with the development of cell-mediated immunity and activation of macrophages to resist intracellular bacterial replication. Both phenomena are controlled by the production cytokines, which occur during infection. Among these cytokines, gamma interferon is a macrophage-activating factor which was shown to activate rodent macrophages to resist *Brucella in vitro* or *in vivo*.[16] Of particular interest is the observation that alcohol use decreases Th1 cytokine levels and responses, and increases Th2 cytokine levels.[17] Therefore, ethanol predominantly impairs

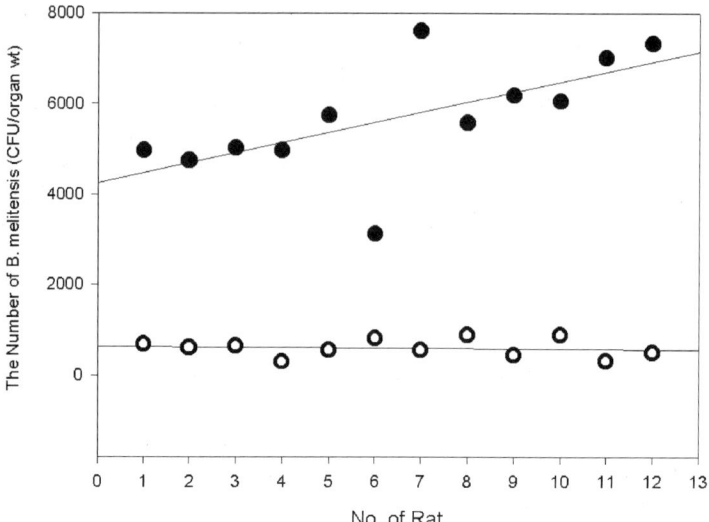

FIGURE 1. Eltanol-receiving rats exposed to *B. melcitensis* infection.

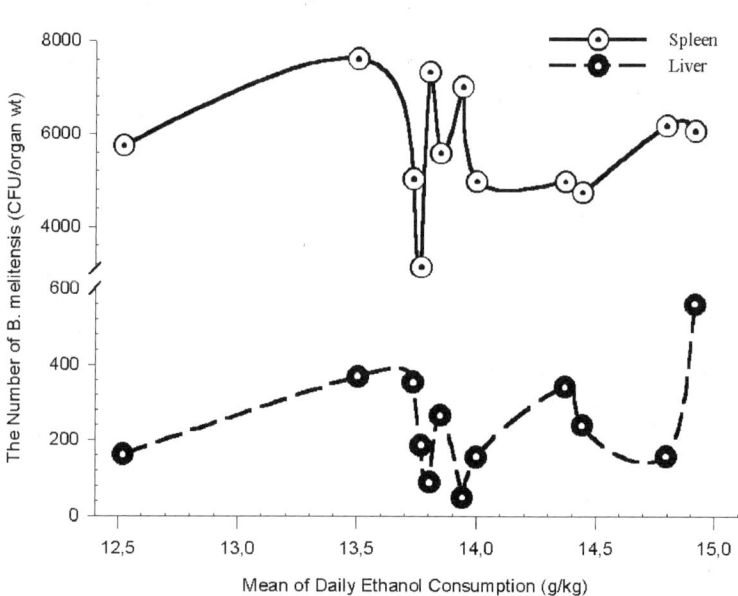

FIGURE 2. Relationship between the amount of ethonol consumption and the *B. melitensis* in spleen.

the ability of mononuclear phagocytes to control the growth of the intracellular organisms.[18] The increase of *Brucella* numbers in the host is mainly due to their ability to avoid the killing mechanisms and proliferate within macrophages like other intracellular pathogens. It has also been shown that the neutrophils of alcohol-fed rats phagocytosed bacteria efficiently, but they do not kill all strains of pneumonia-causing bacteria with normal effectiveness.[19,20]

The animal model, which mimics human brucellosis, was developed and used to study the efficacy of various antibiotics in the treatment. The criteria for therapeutic efficacy in brucellosis animal models are: cure documented by the sterilization of the animals' spleen or reduction of viable counts of *Brucella* cultured from the homogenized spleens.[21] The studies employed the same model using mice. In the study of *Brucella* infection and ethanol, the same basic model was followed with the exception that rats were used instead of mice. According to some authors, reduction in spleen weights is also a criterion for the therapeutic efficacy. In contradiction to the findings of Philippon and coworkers,[22] the difference in experimental animals may account for the fact that weight loss and massive splenomegaly were not prominent features of disease activity in the control mice reported from other studies.[8,21,23] Therefore, the insignificant difference that is found for the spleen and liver—body ratio between groups of *Brucella* infection controls and those given ethanol—does not rule out the severity of infection in ethanol-treated rats.

The *Brucella* infection and ethanol model of rats offer a means to study host response to infection under controlled conditions so that associated risk factors can be assessed and modulating influences (i.e., malnutrition, gender difference) can be studied.

3. EFFICACY OF COMBINATIONS OF DOXYCYCLINE AND RIFAMPICIN

The *Brucella* infection and ethanol model of rats was followed with adding brucellosis treatment procedures. The efficacy of combination of doxycycline and rifampicin in the therapy of ethanol-treated rats was evaluated. The drugs were administered intragastrically starting on day 7 following *B. melitensis* inoculation. Antibiotic dosage was selected on the basis of previous experimental data.[8,21] in which 6 mg/kg/day of rifampicin and 10 mg/kg/day doxycycline were shown to cause a complete cure when administered singly and a linear dose response cure was demonstrated.

A comparison of rat weights, spleen weights, and the ratio of the two did not reveal any consistent pattern. These data are not addressed as parameters of therapeutic outcome. The ranges of blood antibiotic levels obtained at various times in random samplings during antibiotic therapy varied. During therapy, maximal, minimal, and mean blood antibiotic levels in relation to the respective MICs for drugs were consistently high. Blood antibiotic levels for both agents were sustained at therapeutic levels (24 hr for doxycycline and 48 hr for rifampicin) following a single intragastric administration.

A combination of doxycycline (10 mg/kg/day) rifampicin (6 mg/kg/ day) administered intragastrically for days sterilized 64.7% of the spleen of alcoholic rats (unpublished results). Despite the availability of many antibacterial agents, the complete cure of infection with prevention of frequent relapses is still an unattainable goal. The treatment of brucellosis complication such as meningitis and endocarditis pose special problems, and there is no unanimity of opinion regarding the optimal regimen. No satisfactory vaccines against human brucellosis are available and worldwide brucellosis remains a major source of disease in humans and domesticated animals.[2] In humans, the disease is severe, and without effective treatment, the disease might lead to a fatal outcome. Since the clinical picture in human brucellosis can be misleading and ethanol abuse is a serious health problem, much attention should be paid to these two intersecting groups.

To eradicate human brucellosis, control of brucellosis in agricultural animals is crucial because of the zoonotic aspects of this infection. Current knowledge on the dissemination of brucellosis considers human-to-human transmission insignificant. However, because of the significances of human transmission, the hope of eradication of brucellosis is problematic. Currently mice are used as animal models for brucellosis and molecular genetic tools for *Brucella* are available, providing efficient experimental tools to investigate bacterial pathogenesis and host immune response for the development of vaccine strains. Novel *Brucella* vaccine strains must be made based on an insightful understanding of bacterial pathogenesis and host immunity. Therefore, generation of live, attenuated strains based on host–pathogen interactions is being explored in combination with rapid and effective methods to detect attenuation in *Brucella* strains.[24]

Further research will undoubtedly develop improved diagnostic methods, immunizing agents, and treatment regimens. Consideration of immunodeficiency state such as alcoholism in the treatment of brucellosis may decrease the prevalence of complications and lead to effective treatment regimens.

REFERENCES

1. Yumuk, Z., Ozdemirci, S., Erden, B. F., and Dundar, V. (2001). The effect of long-term ethanol feeding on *Brucella melitensis* infection of rats. *Alcohol Alcohol.* **36**:314–317.
2. Corbel, M. J. (1997). Brucellosis: An overview. *Emerg. Infect. Dis.* **3**:213–221.
3. Franz, D. R., Jahrling, P. B., Friedlander, A. M., McClain, D. J., Hoover, D. L., Bryne, W. R. *et al.* (1997). Clinical recognition and management of patients exposed to biological warfare agents. *JAMA* **278**:399–411.
4. Verger, J. M., Grimont, F., Grimont, P. D. A., and Grayon, M. (1985). Brucella, a monospecific genus as shown by deoxyribonucleic acid hybridization. *Int. J. Syst. Bacteriol.* **35**:292–295.
5. Boschiroli, M. L., Foulongne, V., and O'Callaghan, D. (2001). Brucellosis: A worldwide zoonosis. *Curr. Opin. Microbiol.* **4**:58–64.
6. Young, E. J. (1995). Brucella species. In G. L. Mandell, J. E. Bennett, and R. Dolin (eds), *Principles and Practice of Infectious Disease*, Churchill Livingstone, New York, NY 10011, pp. 2053–2060.

7. Kocagoz, S., Akova, M., Altun, B., Gur, D., and Hascelik, G. (2002). In vitro activities of new quinolones against *Brucella melitensis* isolated in a tertiary-care hospital in Turkey. *Clin. Microbiol. Infect.* **8:**240–242.

8. Shasha, B., Lang, R., and Rubinstein, E. (1992). Therapy of experimental murine brucellosis with streptomycin, co-trimoxazole, ciprofloxacin, ofloxacin, pefloxacin, doxycycline, and rifampin. *Antimicrob. Agents Chemother.* **36:**973–976.

9. Bagasra, O., Howeedy, A., Dorio, R., and Kajdacsy-Balla, A. (1987). Functional analysis of T-cell subsets in chronic experimental alcoholism. *Immunology* **61:**63–69.

10. Thiele, G. M., Szabo, G., Kovacs, E. J., Bautista, A. P., Sosa, L., and Jerrells, T. R. (2002). Modulation of immunity and viral–host interactions by alcohol. *Alcohol Clin. Exp. Res.* **26:**1897–1908.

11. Sternbach, G. L. (1990). Infections in alcoholic patients. *Emerg. Med. Clin. North Am.* **8:**793–803.

12. Saad, A. J., Domiati-Saad, R., and Jerrells, T. R. (1993). Ethanol ingestion increases susceptibility of mice to Listeria monocytogenes. *Alcohol Clin. Exp. Res.* **17:**75–85.

13. Davis, C. C., Mellencamp, M. A., and Preheim, L. C. (1991). A model of pneumococcal pneumonia in chronically intoxicated rats. *J. Infect. Dis.* **163:**799–805.

14. Yamamoto, Y., Klein, T. W., and Friedman, H. (1993). Differential effects of ethanol on permissive versus nonpermissive macrophages infected with *Legionella pneumophila. Proc. Soc. Exp. Biol. Med.* **203:**323–327.

15. Bermudez, L. E. and Young, L. S. (1985). Ethanol augments intracellular survival of *Mycobacterium avium* complex and impairs macrophage responses to cytokines. *J. Infect. Dis.* **163:**1286–1292.

16. Dornand, J., Gross, A., Lafont, V., Liautard, J., Oliaro, J., and Liautard, J. P. (2002). The innate immune response against Brucella in humans. *Vet. Microbiol.* **90:**383–394.

17. Friedman, H., Newton, C., and Klein, T. W. (2003). Microbial infections, immunomodulation, and drugs of abuse. *Clin. Microbiol. Rev.* **16:**209–219.

18. Jerrells, T. R. and Sibley, D. (1995). Effects of ethanol on cellular immunity to facultative intracellular bacteria. *Alcohol Clin. Exp. Res.* **19:**11–16.

19. Jareo, P. W., Preheim, L. C., Lister, P. D., and Gentry, M. C. (1995). The effect of ethanol ingestion on killing of *Streptococcus pneumoniae, Staphylococcus aureus* and *Staphylococcus epidermidis* by rat neutrophils. *Alcohol Alcohol.* **30:**311–318.

20. Jareo, P. W., Preheim, L. C., and Gentry, M. C. (1996). Ethanol ingestion impairs neutrophil bactericidal mechanisms against *Streptococcus pneumoniae. Alcohol Clin. Exp. Res.* **20:**1646–1652.

21. Shasha, B., Lang, R., and Rubinstein, E. (1994). Efficacy of combinations of doxycycline and rifampicin in the therapy of experimental mouse brucellosis. *J. Antimicrob. Chemother.* **33:**545–551.

22. Philippon, A. M., Plommet, M. G., Kazmierczak, A., Marly, J. L., and Nevot, P. A. (1977). Rifampin in the treatment of experimental brucellosis in mice and guinea pigs. *J. Infect. Dis.* **136:**481–488.

23. Lang, R., Shasha, B., and Rubinstein, E. (1993). Therapy of experimental murine brucellosis with streptomycin alone and in combination with ciprofloxacin, doxycycline, and rifampin. *Antimicrob. Agents Chemother.* **37:**2333–2336.

24. Ko, J. and Splitter, G. A. (2003). Molecular host–pathogen interaction in brucellosis: Current understanding and future approaches to vaccine development for mice and humans. *Clin. Microbiol. Rev.* **16:**65–78.

15

Alcohol, Infection, and the Lung

PING ZHANG, GREGORY J. BAGBY, JAY K. KOLLS,
LEE J. QUINTON, and STEVE NELSON

1. INTRODUCTION

Alcohol abusers are susceptible to a wide range of infectious diseases, particularly pulmonary infections. Factors that contribute to the development of pulmonary infections in alcohol-abusing patients include the loss of protective barriers in the respiratory tract, aspiration of oropharyngeal contents, nutritional deficiencies, liver disease, and inhibition of the immune defense system. In recent years, human immunodeficiency virus (HIV) infection has become epidemic. Individuals, especially young people, who abuse alcohol and other substances, are at significant risk for HIV infection. Studies have shown that as many as 82% of HIV-infected individuals consume alcohol, with 41% classified as alcoholics.[1] Alcohol may increase susceptibility to HIV infection and/or disease progression. Alternatively, the immune system of the alcohol-consuming patient may become further compromised by HIV infection. This chapter discusses the complex host–pathogen interactions in the airways with an emphasis on

PING ZHANG • Department of Medicine, Section of Pulmonary and Critical Care Medicine, and Alcohol Research Center, Louisiana State University Health Sciences Center, New Orleans, LA 70112.   GREGORY J. BAGBY and STEVE NELSON • Department of Medicine, Section of Pulmonary and Critical Care Medicine, Department of Physiology, and Alcohol Research Center, Louisiana State University Health Sciences Center, New Orleans, LA 70112.   JAY K. KOLLS • Department of Medicine, Section of Pulmonary and Critical Care Medicine, Alcohol Research Center and Gene Therapy Programs, Louisiana State University Health Sciences Center, New Orleans, LA 70112.   LEE J. QUINTON • Department of Physiology and Alcohol Research Center, Louisiana State University Health Sciences Center, New Orleans, LA 70112.

Infectious Diseases and Substance Abuse, edited by Herman Friedman *et al.*
Springer, New York, 2005.

how alcohol consumption adversely affects immune defense mechanisms and predisposes the host to infections. New immunomodulatory strategies for improving host defense function in alcoholic patients will also be discussed.

2. EPIDEMIOLOGY AND CLINICAL MANIFESTATIONS

Benjamin Rush, the first Surgeon General of the United States, published *An Inquiry Into the Effects of Ardent Spirits Upon the Human Body and Mind* in 1785, in which he noted that alcoholics were vulnerable to yellow fever, tuberculosis, pneumonia, and abscesses.[2] In 1905, Sir William Osler remarked that alcohol abuse was "perhaps the most potent predisposing factor to lobar pneumonia."[3] Shortly later, Capps and Coleman studied the influence of alcohol on the prognosis of pneumonia in a group of hospitalized patients and showed that the mortality rate of pneumonia was more than twice as high in alcoholics compared to nonalcoholics.[4] Since then, studies have consistently demonstrated that alcohol abuse increases the incidence and severity of pulmonary infections. A large number of studies have been carried out in various medical settings and patient populations. One study of 1,722 alcoholic patients in Oslo during the years of 1925–40 reported that the age-specific death rates caused by pneumonia were more than three times greater in alcoholics compared to those in the general population.[5] Similarly, another study of 1,298 patients treated for lobar pneumonia from 1927 to 1935 showed that the mortality rate in alcoholics was approximately twice that of nonalcoholic patients.[6] In 1972, Schmidt and De Lint reported an investigation of 6,478 alcoholic patients treated at the Toronto Clinic of the Addiction Research Foundation during a 14-year period. The mortality rates of pneumonia in alcoholic men and women in this series were 3-fold and 7-fold greater, respectively, in comparison to those in the general population of Ontario.[7] More recently, Fernandez-Sola reported a two-phase study in a group of middle-aged patients.[8] Among the risk factors analyzed, high alcohol intake was the only independent risk factor for community-acquired pneumonia. Alcoholic patients with pneumonia showed more severe clinical symptoms, required a longer duration of intravenous antibiotics and longer hospital stays, had more multilobar involvement and pleural effusions, as well as slower resolution of pulmonary infiltrates. In addition, high alcohol intake was the only prognostic factor for mortality. A cohort study of 23,198 pneumonia patients hospitalized in 1992 has shown that for pneumonia cases with an alcohol-related diagnosis, risk-adjusted hospital charges were higher, length of hospital stay was longer, and intensive care unit use was more frequent.[9] Another recent study on patients with severe community-acquired pneumonia presenting in septic shock reported that alcohol abuse predisposes patients to pulmonary infections with *Pseudomonas aeruginosa* and *Acinetobacter* species, both of which are frequently fatal (82%). Musher *et al.* reported a prospective study analyzing predisposing factors for pneumococcal pneumonia with and without bacteremia.[10] The results showed that although the mean number of predisposing factors was greater among bacteremic patients than nonbacteremic patients, only alcohol consumption was significantly more common in patients with bacteremia. Alcohol abuse has also been shown to be a significant risk factor for hospital-acquired pneumonia. In 2000, Everts and colleagues reported a 1-year prospective study of consecutive patients hospitalized for general medical and surgical diseases.[11] Nosocomial pneumonia developed in 126 patients representing 6.1 per 1,000 admissions. Fourteen patients (11%) died as a consequence of pneumonia. Alcohol excess was identified as one of the most powerful predictors of a fatal outcome by univariate analysis.

Bacterial pneumonias of all types including Gram-positive and Gram-negative, aerobic and anaerobic, as well as mycobacterial infections are more common in alcoholics compared to nonalcoholics. Alcohol abusers are also susceptible to lung infections caused by "atypical pathogens," fungi, and viruses. Clinical features of pulmonary infections in alcoholic patients are similar to those in the general population, except for a younger age of occurrence, more severe symptoms, a higher incidence of complications, more frequent recurrence, greater likelihood of developing infection with resistant pathogens, and poorer outcomes. Alcoholic patients with cirrhosis or bone marrow suppression have the poorest prognosis. Among all bacterial pneumonias, *Streptococcus pneumoniae* has been reported to be the most frequent pathogen in the general population as well as in alcohol abusers. *Hemophilus influenzae* and *Klebsiella pneumoniae* are also frequent pathogens causing pneumonias in alcoholic patients. Pulmonary infection caused by *K. pneumoniae* is usually life threatening and associated with a high frequency of complications and death. Alcoholic patients have been reported to have a high incidence of pulmonary infections with *P. aeruginosa* and *Acinetobacter* species which frequently result in death. Anaerobic lung infections with *Fusobacterium nucleatum, Bacteriodes melaninogenicus,* and *Bacteriodes fragilis* are frequently observed in alcoholic patients.[12] In fact, studies have shown that about 30% of all anaerobic pulmonary infections occur in heavy alcohol consumers.[12,13] Clinical presentations of anaerobic lung infection include simple pneumonitis, necrotizing pneumonia, lung abscess, and empyema.[12]

The incidence of pulmonary tuberculosis is significantly higher in alcohol abusers than in the general population. Alcoholic patients with pulmonary tuberculosis usually have more extensive disease at the time of initial diagnosis. Consistent with the severity of the disease, alcoholics have a higher risk of death during the initial hospitalization. Addiction to alcohol is an independent risk factor for mortality in tuberculosis patients. A recent study of 1,493 tuberculosis patients showed that patients who used alcohol excessively were at increased risk of hospitalization during treatment.[14] Lack of patient compliance is a significant problem for the effective treatment of tuberculosis in these patients. High rates of relapse and the development of multiple drug-resistant strains are common in this population. The HIV epidemic, especially among substance abusers and alcoholics, has played an important role in the worldwide resurgence of tuberculosis during the last two decades.

Opportunistic pathogens including *Pneumocystis carinii* are a common cause of pulmonary infections in immunocompromised hosts. The etiological significance of these opportunistic pathogens in alcohol-abusing patients has come to attention recently because of the high rate of HIV infection in this patient population. *P. carinii* pneumonia usually occurs in patients with compromised cell-mediated immunity and is one of the most common pulmonary complications in HIV-infected individuals. Alcohol is immunosuppressive and exerts adverse effects on cell-mediated immunity. Experimental studies have shown that mice on a chronic alcohol-containing diet have a significantly increased rate (greater than 60% in the alcohol-fed group vs none in control group) of *P. carinii* infection in the lung following an intrapulmonary challenge with this pathogen.[15] In the clinic, *P. carinii* pneumonia has been observed in patients with alcoholic hepatitis and cirrhosis in the absence of HIV infection.[16]

3. HOST DEFENSE MECHANISMS IN THE AIRWAYS

The human respiratory tract possesses a sophisticated defense system which effectively protects the host from invading pathogens. This system includes both innate (nonspecific) and acquired (specific) immune defenses. Innate defense primarily consists of structural defenses, antimicrobial molecules generated in the airways, and phagocytic defenses provided by the resident alveolar macrophages and the polymorphonuclear leukocytes (PMNs) that are recruited into the lung in response to pulmonary infection. Mechanical host defenses include the structural barriers in the respiratory tract and mucociliary blanket lining the surface of the airways. Mucins in the mucociliary fluid trap airborne particles and microorganisms. The mucus containing trapped particles and microbes are propelled to the oropharynx by ciliary movement. Coughing is an important mechanical defense mechanism responsible for clearing secretions from the airways. Particles less than 5 μm in diameter can bypass these defenses and gain access to the alveolar space. This is particularly relevant in the pathogenesis of pulmonary infection as most bacteria and mycobacteria are within this size range. Airways also produce a variety of antimicrobial molecules which either possess direct antimicrobial activity or facilitate the elimination of infectious pathogens by phagocytes. These include lysozyme, complement, immunoglobulin A and G, fibronectin, lactoferrin, transferrin, lipopolysaccharide-(LPS)-binding protein, defensins, cathelicidins, and collectins.

In the terminal airways, alveolar macrophages constitute the first line of phagocytic defense. These cells reside in the alveolar space and are avidly phagocytic. They are responsible for the clearance of small loads of pathogenic organisms to maintain the sterility of the lung. Certain microorganisms, such as *Mycobacterium* spp. and *Legionella* spp., are resistant to the microbicidal activities of alveolar macrophages and are capable of replicating intracellularly. The eradication of these pathogens requires the involvement of other immune defense mechanisms such as cell-mediated immunity. In the event that the invading pathogens are too virulent or the inoculum is too large, alveolar macrophages are capable of generating numerous mediators that orchestrate the recruitment of PMNs from the systemic circulation into the alveolar space. These recruited PMNs provide auxiliary phagocytic defenses and reinforce the immune response against offending pathogens. Alveolar macrophage-derived substances capable of eliciting PMN migration into the airways include chemotactic peptides such as interleukin-8 (IL-8), macrophage inflammatory protein-2 (MIP-2), and other CXC chemokines, complement fragments including C3a and C5a, and arachidonic acid metabolites such as leukotriene B4.

Cytokines are important mediators responsible for communication between alveolar macrophages and other cellular components of the immune system. They can be divided into proinflammatory and anti-inflammatory subgroups, both of which function critically in the regulation of the pulmonary host defense response including the initiation, localization, reinforcement, and ultimate resolution of the response.

Proinflammatory cytokines that play an important role in pulmonary host defense include tumor necrosis factor-α (TNF-α), IL-8, IL-12, MIP-2, granulocyte colony-stimulating factor (G-CSF), and interferon-γ (IFN-γ). TNF-α has been designated as an early-response or "alarm" cytokine. It is rapidly produced by alveolar macrophages following exposure to infectious agents. TNF-α activates phagocyte functional activities and stimulates the release of other cytokines and chemokines by different types of pulmonary cells including alveolar macrophages and epithelial cells. CXC chemokines, including IL-8 and MIP-2, account for the major chemotactic activity in the alveolar space for PMN recruitment. In addition, CXC chemokines enhance PMN activities including the expression of surface receptors, phagocytosis, and generation of reactive oxygen species. IL-12 promotes Th1-type immune responses and enhances cell-mediated immunity against airway infections caused by viruses, mycobacteria, fungi, and parasites. In addition, IL-12 promotes innate immunity in the lung against bacterial pathogens in experimental models of infection. Patients with IL-12 deficiency develop recurrent pneumococcal pneumonia with sepsis and other infections. G-CSF is a lineage-specific hematopoietic growth factor which selectively stimulates the proliferation and maturation of myeloid progenitor cells to PMNs. It plays a critical role in maintaining the normal blood level of PMNs and is responsible for increasing the number of circulating PMNs during infection. G-CSF also enhances the functional activities of PMNs including adhesion molecule expression, chemotaxis, oxygen metabolism, phagocytosis, and intracellular bacterial killing. IFN-γ exerts profound effects on various aspects of host defense against a wide range of pathogens including viruses, bacteria, fungi, and intracellular and extracellular parasites.[17] IFN-γ enhances cytokine and chemokine production by alveolar macrophages and other types of leukocytes. It also stimulates the respiratory burst and release of lysosomal enzymes from PMNs, and actively modulates antigen presentation, cell differentiation, and cytotoxicity of immune effector cells.

In contrast to proinflammatory cytokines, anti-inflammatory cytokines downregulate host immune responses. IL-10 is a representative of these mediators. IL-10 inhibits the production of many proinflammatory cytokines including TNF-α, IL-1β, IFN-γ, IL-12, MIP-2, and MIP-1α. IL-10 also suppresses the functional activities of PMNs.[18] This cytokine may play an important role in adjusting the intensity of the host response to an infection and mediating the resolution of tissue inflammation once the infection is confined.

One unique feature of the pulmonary host defense response is the selective compartmentalization of certain cytokines and chemokines. Experimental studies have shown that intrapulmonary LPS or bacteria induce a rapid increase of TNF-α and MIP-2 in the lung without an increase of these mediators in the systemic circulation.[19–22] Similar observations have been reported in humans. Patients with unilateral pneumonia have a compartmentalized inflammatory response within the infected lung with localized production of TNF-α, IL-1, IL-6, and IL-8.[23,24] We speculate that this selective increase in proinflammatory cytokines is essential for localizing the inflammatory reaction within the infected compartment. Interestingly, not all cytokines are compartmentalized. Intrapulmonary challenge with LPS or bacteria causes increases in G-CSF and cytokine-induced neutrophil chemoattractant (CINC) in both the BAL fluid and systemic circulation in animal models.[21,22,25,26] Whether a cytokine is compartmentalized or not most likely depends on its physiologic function. The increase in serum G-CSF is pivotal for the production of PMNs from the bone marrow that serves to reinforce the host defense response to infection. CINC has been shown to activate PMNs and enhance their response to other chemokines.[27] The decompartmentalization of these cytokines during infection is likely to be an important mechanism by which they are able to reach the appropriate target organ and exert their physiologic functions.

Although very few PMNs exist in the alveolar space in normal conditions, large numbers of PMNs are maintained in the lung vasculature. The marginated pool of PMNs in the pulmonary vasculature constitutes approximately 40% of the body's total PMNs.[28] Following an appropriate stimulus, PMNs migrate into the alveolar space to reinforce phagocytic and bactericidal defenses. Intrapulmonary challenges in animals with either bacteria or LPS elicit a rapid recruitment of PMNs into the lung. By 3–4 hr after the challenge, PMNs may constitute 60–90% of the total cells recovered by BAL. PMN activation also occurs during the recruitment of these cells due to their exposure to a variety of proinflammatory cytokines and mediators contained within the infected compartment. In addition to the ingestion and killing of invading microorganisms, recruited PMNs may participate in the regulation of the local host defense response by producing different cytokines including TNF-α, IL-1β, IL-6, and MIP-2. PMNs also trap and scavenge chemokines in the surrounding environment, which may play an important role in the resolution of the inflammatory response in the lung.[29,30]

The acquired (or specific) immune defense system is well developed in the human lung, and consists of both humoral and cellular immune components. Specific immunity is the major host defense against pathogens that are able to evade the innate immune defense system. Mounting an acquired immune response in the lung involves a complex interplay between antigen-presenting cells or accessory cells (such as dendritic cells and alveolar macrophages) and lymphocytes (T and B lymphocytes). Antigen-presenting cells capture and process antigen. The processed antigens together with class II major histocompatibility complex (MHC) molecules on the antigen-presenting cell surface are then presented to CD4+ T lymphocytes. The activated CD4+ T lymphocytes subsequently develop into specific helper T (T_H) cells to produce various types of cytokines. These cytokines play an essential role in mediating the proliferation and activation of immune effector cells including B lymphocytes and cytotoxic T lymphocytes (CTL). The pattern of cytokines produced by the T_H cells determines the degree to which the humoral or cell-mediated branches of the immune system are predominantly activated. In the lung, the major antigen-presenting cells are dendritic cells. Alveolar macrophages are weak at presenting antigens. It has been proposed that alveolar macrophages may play a role in "antigen transfer." Alveolar macrophages initially take up the antigen and then transfer the processed peptides to dendritic cells for efficient presentation.[31] Under certain conditions however, such as HIV infection, alveolar macrophages have been shown to have an enhanced activity in stimulating T-cell proliferation.[32] Antigens in the alveolar space may either directly diffuse into regional lymphoid tissues or be captured by antigen-presenting cells which then migrate to regional lymph nodes. Within these regional lymphoid tissues, the primary immune response is initiated and a large number of immune effector cells including CTLs and antibody producing B cells are produced. The generated effector B and T cells traffic back to the lung through the systemic circulation and eventually reside in the interstitium and alveolar space by means of their homing mechanisms. Mounting a specific immune response to a new antigen takes place over a period of days to weeks. Memory B and T cells are also created during this process. These memory cells can rapidly (hours to days) organize a response when the host is subsequently exposed to the same antigen.[31] Memory cells constitute the predominant type of lymphocyte residing in normal lungs.[33]

4. ALCOHOL AND PULMONARY HOST DEFENSE

Alcohol consumption impairs both innate and acquired immunity. Inhibition of neutrophil function is one of the most extensively characterized immune defects induced by alcohol. In 1938, Pickrell observed that rabbits intoxicated with alcohol failed to mount an acute leukocytic response to pneumococcal infection in the lung.[34] Since then studies on experimental animals and human subjects have repeatedly shown that alcohol blocks tissue delivery of PMNs during infection and inflammation. In 1964, Green and colleagues documented that pulmonary clearance of bacteria was suppressed by alcohol intoxication.[35] Two decades later, Astry and colleagues studied the relationship between the alcohol-induced defects of PMN recruitment and pulmonary clearance of bacteria.[36] In their studies, animals were challenged by aerosol inhalation of either Gram-positive (*Staphylococcus aureus*) or Gram-negative (*Proteus mirabilis*) bacteria in the presence and absence of acute alcohol intoxication. Alcohol caused a dose-dependent inhibition of PMN recruitment into the alveolar space following the bacterial challenge. In association with this impaired PMN influx, pulmonary clearance of both the Gram-positive and Gram-negative bacteria was suppressed by alcohol in a dose-dependent manner. Similar observations have been reported in various experimental models with intrapulmonary challenges of different pathogens.

PMN recruitment into tissue sites of infection and inflammation is a complex process involving the margination, adhesion, and transendothelial migration of these phagocytes. An intricate interplay of various adhesion molecules on the surface of both PMNs and the endothelium of the microvasculature takes place during this multistep process. A high concentration of chemoattractants produced in the inflammatory tissue is pivotal in guiding PMN migration from the vasculature into the infected site. Alcohol has been known to exert inhibitory effects on several steps in this process. Normally, the expression of β_2-integrin adhesion molecules CD11b/CD18 on PMNs is rapidly upregulated upon activation. CD11b/CD18 mediates PMN firm attachment to the endothelium and their subsequent transendothelial migration. Alcohol inhibits upregulation of CD18 expression on PMNs in response to inflammatory stimuli[37] and suppresses PMN "hyperadherence" to endothelial monolayers following appropriate stimulation.[38] Our studies have shown that alcohol intoxication suppresses the upregulation of CD11b/c and CD18 expression on circulating PMNs in animals challenged with systemic LPS.[39] Other investigators have also reported that alcohol causes a dose-dependent decrease in granulocyte adherence, which correlates with the observed inhibition of PMN tissue delivery.

A normal PMN response to chemoattractants is essential for the directed migration of these phagocytes. Alcohol has been shown to inhibit the PMN response to chemoattractants. Experimental studies show that administration of alcohol to rats results in a significant decrease in PMN chemotaxis to LPS-activated normal rat serum.[40] PMNs from individuals who abuse alcohol also exhibit a decreased chemotactic response. In patients with alcoholic cirrhosis, LPS absorbed from the portal system may gain access to the systemic circulation due to either the development of a shunt between these two systems or impaired Kupffer cell function. This "spillover" of LPS into the systemic circulation may induce a chronic inflammatory reaction in the host. Chemoattractants such as CXC chemokines (IL-8) and complement fragments (C5a) are elevated in the peripheral circulation of patients with alcoholic liver disease. The chronic *in vivo* activation of PMNs has been postulated to account for the blunted response of PMNs to chemoattractants in these hosts.

In contrast to the events that occur in chronic alcoholic patients, acute alcohol intoxication causes a profound inhibition of CXC chemokine production in the lung during pulmonary infection and inflammation.[20,41] This inhibition occurs at the level of both gene expression and protein production. Insufficient production of chemokines in the alveolar space diminishes the chemotactic gradient across the alveolar-capillary membrane. Thus, the signals that trigger PMN migration into the infected focus are reduced.

PMN release from the hematopoietic tissue (bone marrow) in response to bacterial infection is an important mechanism for recruitment of additional phagocytic cells. Neither acute nor chronic drinking in a controlled environment affects PMN release from the bone marrow in response to appropriate stimulation.[42] G-CSF stimulates myeloid progenitor cell proliferation to PMNs and the release of PMNs from bone marrow to the peripheral circulation. Certain CXC chemokines including IL-8 and MIP-2 may also promote bone marrow granulopoiesis and the release of granulocytes. We, and others, have shown that G-CSF and CXC chemokine levels in the peripheral circulation increase significantly during pulmonary infections. Alcohol intoxication suppresses both the G-CSF and chemokine responses in experimental animals challenged with either pulmonary or systemic bacterial pathogens.[20,43,44] Clinical investigations have shown that a significant number of hospitalized alcohol-abusing patients with infections present with granulocytopenia at admission, which is a predictor of increased mortality.[45,46] Bone marrow examination has shown a significant reduction in the number of mature granulocytes with vacuolization of myeloid progenitor cells in alcohol-abusing patients. Incubation of bone marrow cells with alcohol at concentrations commonly observed in intoxicated patients has been reported to suppress granulocyte colony formation.[47,48]

PMN functional activities are also affected by alcohol. In addition to the inhibition of adhesion molecule expression and adherence of PMNs as mentioned previously, *in vitro* studies have shown that high concentration of alcohol (>640 mg%) inhibits human PMN phagocytosis and intracellular killing of *S. aureus*.[49] Alcohol at clinically relevant levels inhibits fMLP-stimulated superoxide production by human PMNs in a dose-dependent manner. Degranulation (elastase release) and bactericidal activity (killing of *S. aureus*) of human PMNs are also inhibited by alcohol at concentrations between 0.2% and 0.3%.[50] *In vivo* intoxication of animals with acute alcohol (blood alcohol concentration of 50–100 mM) results in a significant inhibition of PMN phagocytic activity[41,51]. PMNs from alcohol intoxicated patients are reported to contain 31% less elastase activity compared to those from normal individuals and produce 25–27% less superoxide than controls in response to inflammatory stimuli.[52]

As described above, alveolar macrophages are the resident phagocytic cells that respond to infectious challenges in the terminal airways. Activated alveolar macrophages produce large amounts of TNF, which serves as a key step in triggering the inflammatory response in the lung. Acute alcohol intoxication suppresses the pulmonary TNF response to bacterial challenges which is associated with an inhibition of PMN recruitment into the alveolar space and clearance of bacteria from the airways. This alcohol-induced inhibition of TNF production by alveolar macrophages occurs at a post-transcriptional level. In alveolar macrophages recovered from rhesus macaques incubated with alcohol (100 mM) 30 min before LPS stimulation, alcohol suppressed LPS-induced TNF protein production by 84% and 70% at 2 and 8 hr, respectively, without affecting the upregulation of TNF mRNA expression by these macrophages.[53] Exposure of monocytes/macrophages to alcohol causes a significant increase in cell-associated TNF in these cells following LPS stimulation.[54,55] These studies suggest that alcohol may impair mechanisms involved in the release of TNF from these cells. In addition to a direct inhibition of proinflammatory cytokine production, alcohol may increase anti-inflammatory cytokine (IL-10) expression by human monocytes, which has been postulated to be one mechanism underlying the immunosuppressive effects of alcohol.[56] At the present time, it remains to be determined whether alcohol exerts the same effect on alveolar macrophage production of IL-10 and, thereby, modulates the pulmonary host defense response.

Macrophage mobilization, adherence, phagocytosis, superoxide production, and microbicidal activity are inhibited by alcohol. These alcohol-induced defects of alveolar macrophage function diminish the capacity of these cells to contain invading pathogens within the alveolar space. This effect of alcohol may be of particular importance in tuberculosis where greater than 90% of inhaled mycobacteria are normally ingested and destroyed by alveolar macrophages.[57] The initial interaction of alveolar macrophages with this pathogen is critical for eliminating the infection. Tubercle bacilli not killed by alveolar macrophages survive and proliferate intracellularly. Studies have shown that exposure to alcohol enhances intracellular growth of mycobacteria in human macrophages.

Alcohol consumption has also been shown to suppress acquired immune defenses including both cell-mediated and humoral immunities. The ability to develop delayed hypersensitivity skin test reactions to various antigens is usually poor in alcohol-abusing patients. Exposure of human monocytes to alcohol suppresses their capacity to present antigen to T cells. Defective antigen presentation has also been observed in animals fed an alcohol-containing diet.[58] Chronic alcohol abusers, especially those with liver disease, frequently develop lymphopenia. Alcohol also suppresses lymphocyte blast transformation in response to mitogen stimulation. Lymphocyte proliferative responses to specific antibodies against T-cell receptors are blunted by alcohol.[59] Chronic alcohol feeding results in atrophy of the thymus and spleen in experimental animals. Chronic alcohol intoxication causes a significant reduction in absolute numbers of CD4+ T lymphocytes in experimental animals. In addition, T lymphocytes isolated from alcoholic hosts have a diminished capacity to produce IFN-γ, an important cytokine that stimulates cell-mediated immunity.[60] Pulmonary recruitment of both CD4+ and CD8+ T lymphocytes in response to *P. carinii* infection in the lung is suppressed by alcohol consumption.[61,62] An increase in plasma immunoglobulins has been observed in alcohol-abusing patients, especially those with alcoholic liver disease. These immunoglobulins do not appear to be protective. Interestingly, the ability of developing specific antibodies following new antigen challenges is impaired in animals chronically intoxicated with alcohol. Since specific antibodies are important in protecting the host against infections caused by certain bacterial pathogens, such as *Streptococcus pneumoniae*, this defect may adversely affect the eradication of these pathogens in patients with pneumonia.

In recent years, HIV infection has become a major public health problem. Studies have shown a significant association between alcohol consumption and the risk of being infected with HIV. One possible explanation for this association is that alcohol consumption increases the likelihood of risky sexual behavior or the incidence of exposure to HIV. At the present time, it remains to be defined whether alcohol also increases the likelihood of disease transmission in individuals exposed to HIV. Bagasra and colleagues reported that alcohol administration in HIV seronegative humans significantly increases HIV replication in peripheral blood mononuclear cells (PBMC) when these cells are infected *in vitro*.[63] Reduction in CD8+ lymphocytes secondary to excessive alcohol consumption may serve as one mechanism underlying this enhanced viral replication.[64] Saravolatz *et al.* have shown that alcohol added to CEM cells prior to exposure to HIV stimulates production of p24 antigen.[65] Alcohol has also been documented to selectively impair the *in vitro* antigenic proliferative response to

HIV env-gag peptide and natural killer cell activity by lymphocytes obtained from AIDS patients.[66]

HIV disease progression has been reported to be adversely affected by alcohol consumption. Fong *et al.* described an alcohol-abusing patient who developed accelerated disease progression to AIDS over a 3-month period.[67] In addition, it has been shown that blood CD4+ lymphocyte counts are increased in alcohol-abusing HIV infected patients during alcohol withdrawal.[68] The simian immunodeficiency virus (SIV) macaque model of HIV infection provides an excellent means to study the interactions of alcohol and SIV infection in a controlled and monitored environment. SIV consists of a group of lentiviruses that are structurally, biologically, antigenically, and genetically related to HIV. SIV has a similar tropism to infect CD4+ cells resulting in acquired immunodeficiency that progresses to AIDS with the occurrence of opportunistic infections. We have conducted a longitudinal study of 32 rhesus macaques. Chronic binge administration of alcohol has been induced in these animals by infusing alcohol or iso-caloric sucrose via a surgically implanted gastric catheter for 5 hr per day for four consecutive days per week with a target blood alcohol concentration of 50–60 mM. In these animals, the plasma viral loads attained between 60 and 120 days postinfection are significantly higher in the alcohol treatment group. As in humans infected with HIV, several studies have established the importance of viral load during this period as a prognostic indicator of disease in the SIV-macaque model.[69,70] Plasma viremia can be considered a crude reflection of the overall level of viral replication in tissues in anatomic continuity with the plasma compartment. The viral load at this point of time postinfection is known as the "viral set point" and is reported to be the most reliable predictor of disease outcome in the SIV-macaque model. These observations suggest that alcohol consumption alters host-HIV interactions which may accelerate disease progression. Pulmonary infections are among the most frequent complications in AIDS patients. Studies examining the sequelae of bacterial pneumonia in our SIV-infected alcoholic rhesus model are currently underway in our laboratory. A better understanding of the impact of alcohol abuse on HIV infection and its relation to pulmonary immune defense will improve our knowledge about the pathogenesis of infectious complications in the respiratory tract in these hosts.

5. IMMUNOMODULATION AND TREATMENT OF PULMONARY INFECTIONS

At the present time, treatment of pulmonary infections in both alcoholic and nonalcoholic patients primarily depends on antibiotic therapy. Since the emergence of drug-resistant pathogens, antibiotic therapy is becoming more problematic. Immunomodulation may be useful as adjuvant therapy in managing pulmonary infections in alcoholics.

Alcohol inhibits the pulmonary innate immune response, especially the recruitment of PMNs into the terminal airways, which is a major risk factor for bacterial pneumonia. Strategies have been developed to augment pulmonary phagocytic defenses either by increasing the number and function of circulating PMNs or enhancing chemotactic signals for PMN migration and activation in the lung. Studies have shown that administration of exogenous G-CSF stimulates PMN release from the bone marrow and augments PMN recruitment into the lung in response to infectious stimuli. Subcutaneous injections of G-CSF (50 μg/kg) twice daily for 2 days results in a 7-fold increase in circulating PMNs and 5-fold increase in PMN influx into the alveolar space in rats following an intratracheal LPS challenge.[41] This enhanced PMN recruitment is not solely driven by the increased number of circulating PMNs. G-CSF treatment also upregulates PMN sensitivity to chemotactic signals.[29] Enhancement of pulmonary antibacterial defenses by G-CSF has been shown in rats infected with *K. pneumoniae.*[71] In these experiments, G-CSF augmented the pulmonary recruitment of PMNs in infected control rats and significantly attenuated the adverse effects of ethanol on PMN delivery into the infected lung. G-CSF also enhanced the pulmonary clearance of bacteria in both control and ethanol-treated rats and improved the survival of these animals. G-CSF has been shown to attenuate the adverse effects of alcohol on many PMN functions, including the expression of adhesion molecules and phagocytosis.[41,72,73]

In a clinical trial of 756 patients with community-acquired pneumonia, subcutaneous injection of G-CSF (300 μg/day) to patients for up to 10 days caused a 3-fold increase in the number of circulating PMNs.[74] G-CSF treatment was well-tolerated by these patients. A faster resolution of X-ray abnormalities and fewer complications including the adult respiratory distress syndrome and disseminated intravascular coagulation were observed in patients treated with G-CSF. These clinical observations suggest that G-CSF may be useful in combination with antibiotics for the treatment of pulmonary infections in patients immunocompromised by alcohol. However, further studies are needed to support this indication.

IFN-γ produced by T lymphocytes was initially identified as a peptide with antiviral and antitumor activities. It is now known that IFN-γ enhances host defense against a wide profile of pathogenic organisms including viruses, bacteria, fungi, and parasites. *In vitro* studies have shown that macrophages stimulated by IFN-γ are able to kill over three dozen different pathogens.[75] IFN-γ administration in conjunction with antibiotic therapy produces synergistic or additive effects in the treatment of certain pathogens (*S. aureus, P. carinii,* and *Cryptococcus neoformans*) that cause lung infections in immunocompromised hosts. Intratracheal instillation or aerosol inhalation of IFN-γ results in activation of alveolar macrophages and augmentation of pulmonary microbicidal activities.[28,76–78] We showed that administration of a recombinant adenoviral vector encoding the murine IFN-γ complementary DNA to rat lung produced prolonged expression of biologically active IFN-γ in the lung. Pulmonary TNF production, PMN recruitment, and bactericidal activity were significantly enhanced in these animals in both normal and alcohol-intoxicated hosts.[79,80] Our recent studies have shown that intratracheal administration of IFN-γ to rats markedly enhances the pulmonary CXC chemokine response to a subsequent LPS challenge.[81] IFN-γ treatment also attenuates acute alcohol-induced suppression of MIP-2 and CINC production in the lung following an intrapulmonary LPS challenge.

Clinical trials of IFN-γ therapy in alcohol-abusing patients with infections have not yet been undertaken. Previous studies have shown that IFN-γ administered either locally or systemically for the treatment of pulmonary and other infections is well tolerated by patients. In patients with disseminated atypical mycobacterial infection (*M. avium complex*), IFN-γ treatment in combination with anti-mycobacterial chemotherapy results in clinical improvement. The treated patients rapidly cleared the infection and became afebrile.[82] Similar results have been seen in patients with AIDS. In a clinical study of patients with multidrug-resistant tuberculosis, aerosol administration of 500 μg IFN-γ three times a week for 1 month eradicated mycobacteria in sputum in all patients.[83] Based on preliminary animal studies and the clinical data to date, such an approach may be of benefit in patients immunocompromised by alcohol.

6. PERSPECTIVE

The function of the pulmonary host defense system is significantly compromised by alcohol. This defect leads to an increased risk for developing a wide spectrum of pulmonary infections. Bacterial pneumonia and other lung infections are more common and severe in individuals who abuse alcohol. Treatment of these infections is usually problematic. Aggressive antimicrobial regimens, in conjunction with immunotherapy, may provide a new approach in the management of these infections in immunocompromised patients.

ACKNOWLEDGMENTS. This work was supported by NIH grant AA-09803 and Louisiana State HEF Grant 2000-05-06.

REFERENCES

1. Lefevre, F., O'Leary, B., Moran, M., Mossar, M., Yarnold, P. R., Martin, G.J. *et al.* (1995). Alcohol consumption among HIV-infected patients. *J. Gen. Intern. Med.* **10:**458–460.
2. Rush, B. (1943). An inquiry into the effects of ardent spirits upon the human body and mind. *Q. J. Stud. Alcohol* **4:**321–341.
3. Olser, W. (1905). *The Principles and Practice of Medicine.* Appleton, New York.
4. Capps, J. A. and Coleman, G. H. (1923). Influence of alcohol on prognosis of pneumonia in Cook County Hospital. *JAMA* **80:**750–752.
5. Sundby, P. (1976). *Alcoholism and mortality.* Oslo, Universitetsforlaget, National Institute for Alcohol Research, Publ. No. 6.
6. Painton, J. F. and Ulrich, H. J. (1937). Lobar pneumonia; an analysis of 1298 cases. *Ann. Intern. Med.* **10:**1345–1364.
7. Schmidt, W. and De Lint, J. (1972). Causes of death of alcoholics. *Q. J. Stud. Alcohol* **33:**171–185.
8. Fernandez-Sola, J., Junque, A., Estruch, R., Monforte, R., Torres, A., and Urbano-Marquez, A. (1995). High alcohol intake as a risk and prognostic factor for community-acquired pneumonia. *Arch. Intern. Med.* **155:**1649–1654.
9. Saitz, R., Ghali, W. A., and Moskowitz, M. A. (1997). The impact of alcohol-related diagnoses on pneumonia outcomes. *Arch. Intern. Med.* **157:**1446–1452.

10. Musher, D. M., Alexandraki, I., Graviss, E. A., Yanbeiy, N., Eid, A., Inderias, L. A. *et al.* (2000). Bacteremic and nonbacteremic pneumococcal pneumonia, A prospective study. *Medicine* **79:**210–221.

11. Everts, R. J., Murdoch, D. R., Chambers, S. T., Town, G. I., Withington, S. G., Martin, I. R. *et al.* (2000). Nosocomial pneumonia in adult general medical and surgical patients at Christchurch Hospital. *NZ Med. J.* **113:**221–224.

12. Bartlett, J. G. and Finegold, S. M. (1974). Anaerobic infections of the lung and pleural space. *Am. Rev. Respir. Dis.* **110:**56–77.

13. Kharkar, R. A. and Ayyar, V. B. (1981). Aetiological aspects of lung abscess. *J. Postgrad. Med.* **27:**163–166.

14. Taylor, Z., Marks, S. M., Rios Burrows, N. M., Weis, S. E., Stricof, R. L., and Miller, B. (2000). Causes and costs of hospitalization of tuberculosis patients in the United States. *Int. J. Tuber. Lung Dis.* **4:**931–939.

15. D'Souza, N. B., Mandujano, J. F., Nelson, S., Summer, W. R., and Shellito, J. E. (1995). Alcohol ingestion impairs host defenses predisposing otherwise healthy mice to *Pneumocystis carinii* infection. *Alcohol Clin. Exp. Res.* **19:**1219–1225.

16. Ikawa, H., Hayashi, Y., Ohbayashi, C., Tankawa, H., and Itoh, H. (2001). Autopsy case of alcoholic hepatitis and cirrhosis treated with corticosteroids and affected by *Pneumocystis carinii* and cytomegalovirus pneumonia. *Pathol. Int.* **51:**629–632.

17. Murray, H. W. (1996). Current and future clinical applications of interferon-gamma in host antimicrobial defense. *Intensive Care Med.* **22:**S456–S461.

18. Laichalk, L. L., Danforth, J. M., and Standiford, T. J. (1996). Interleukin-10 inhibits neutrophil phagocytic and bactericidal activity. *FEMS Immunol. Med. Microbiol.* **15:**181–187.

19. Nelson, S., Bagby, G. J., Bainton, B. G., Wilson, L. A., Tompson, J. J., and Summer, W. R. (1989). Compartmentalization of intraalveolar and systemic lipopolysaccharide-induced tumor necrosis factor and the pulmonary inflammatory response. *J. Infect. Dis.* **159:**189–194.

20. Boé, D. M., Nelson, S., Zhang, P., and Bagby, G. J. (2001). Acute ethanol intoxication suppresses lung chemokine production following infection with *Streptococcus pneumoniae. J. Infect. Dis.* **184:**1134–1142.

21. Quinton, L. J., Nelson, S., Boé, D. M., Zhang, P., Zhong, Q., Kolls, J. K. *et al.* (2002). The granulocyte colony-stimulating factor response following intrapulmonary and systemic bacterial challenges. *J. Infect. Dis.* **185:**1476–1482.

22. Zhang, P., Nelson, S., Holmes, M. C., Summer, W. R., and Bagby, G. J. (2002). Compartmentalization of macrophage inflammatory protein-2, but not cytokine-induced neutrophil chemoattractant, in rats challenged with intratracheal endotoxin. *Shock* **17:**104–108.

23. Dehoux, M. S., Boutten, A., Ostinelli, J., Seta, N., Dombret, M. C., Crestani, B. *et al.* (1994). Compartmentalized cytokine production within the human lung in unilateral pneumonia. *Am. J. Respir. Crit. Care Med.* **150:**710–716.

24. Boutten, A., Dehoux, M. S., Seta, N., Ostinelli, J., Venembre, P., Crestani, B. *et al.* (1996). Compartmentalized IL-8 and elastase release within the human lung in unilateral pneumonia. *Am. J. Respir. Crit. Care Med.* **153:**336–342.

25. Nelson, S., Zhang, P., Summer, W., and Bagby, G. (2000). Granulocyte colony-stimulating factor is not compartmentalized following an intrapulmonary bacterial challenge. *Am. J. Respir. Crit. Care Med.* **161:**A216.

26. Blackwell, T. S., Lancaster, L. H., Blackwell, T. B., Venkatakrishnan, A., and Christman, J. W. (1999). Chemotactic gradients predict neutrophilic alveolitis in endotoxin-treated rats. *Am. J. Respir. Crit. Care Med.* **159:**1644–1652.

27. Shibata, F., Konishi, K., Kato, H., Komorita, N., Al-Mokdad, M., Fujioka, M. *et al.* (1995). Recombinant production and biological properties of rat cytokine-induced neutrophil chemoattractants, gro/CINC-2α, CINC-2β, and CINC-3. *Eur. J. Biochem.* **321:**306–311.

28. Nelson, S., Mason, C. M., Kolls, J., and Summer, W. R. (1995). Pathophysiology of pneumonia. *Clin. Chest Med.* **16:**1–12.

29. Zhang, P., Bagby, G. J., Kolls, J. K., Welsh, D. A., Summer, W. R., Andresen, J. *et al.* (2001). The effects of granulocyte colony-stimulating factor and neutrophil recruitment on the pulmonary chemokine response to intratracheal endotoxin. *J. Immunol.* **165:**458–468.

30. Holmes, M. C., Zhang, P., Nelson, S., Summer, W. R., and Bagby, G. J. (2002). Neutrophil modulation of the pulmonary chemokine response to lipopolysaccharide. *Shock* **18:**555–560.

31. Twigg III, P. H. (1998). Pulmonary host defenses. *J. Thorac. Imag.* **13:**221–233.

32. Twigg, H. L., Lipscomb, M. F., Yoffe, B., Barbaro, D., and Weissler, J. C. (1989). Enhanced accessory cell function by alveolar macrophages from patients infected with the human immunodeficiency virus: Potential role for depletion of CD4+ T cells in the lung. *Am. J. Respir. Cell Mol. Biol.* **1:**391–400.

33. Saltini, C., Kirby, M., Trapnell, B. C., Tamura, N., and Crystal, R. G. (1990). Biased accumulation of T lymphocytes with "memory"-type CD45 leukocyte common antigen gene expression on the epithelial surface of the human lung. *J. Exp. Med.* **170:**1123–1140.

34. Pickrell, K. L. (1938). The effect of alcoholic intoxication and ether anesthesia on resistance to pneumococcal infection. *Bull. Johns Hopkins Hosp.* **63:**238–260.

35. Green, G. M. and Kass, E. H. (1964). Factors influencing the clearance of bacteria by the lung. *J. Clin. Invest.* **43:**769–776.

36. Astry, C. L., Warr, G. A., and Jakab, G. J. (1983). Impairment of polymorphonuclear leukocyte immigration as a mechanism of alcohol-induced suppression of pulmonary antibacterial defenses. *Am. Rev. Respir. Dis.* **128:**113–117.

37. Nilsson, E., Lindström, P., Patarroyo, M., Ringertz, B., Lerner, R., Rincon, J. *et al.* (1991). Ethanol impairs certain aspects of neutrophil adhesion in vitro: comparisons with inhibition of expression of the CD18 antigen. *J. Infect. Dis.* **163:**591–597.

38. MacGregor, R. R., Safford, M., and Shalit, M. (1988). Effect of ethanol on functions required for the delivery of neutrophils to sites of inflammation. *J. Infect. Dis.* **157:**682–689.

39. Zhang, P., Bagby, G. J., Xie, M., Stoltz, D. A., Summer, W. R., and Nelson, S. (1998). Acute ethanol intoxication inhibits neutrophil β_2-integrin expression in rats during endotoxemia. *Alcohol Clin. Exp. Res.* **22:**135–141.

40. Lister, P. D., Gentry, M. J., and Preheim, L. C. (1993). Ethanol impairs neutrophil chemotaxis *in vitro* but not adherence or recruitment to lungs of rats with experimental pneumococcal pneumonia. *J. Infect. Dis.* **167:**1131–1137.

41. Zhang, P., Bagby, G. J., Stoltz, D. A., Summer, W. R., and Nelson, S. (1999). Granulocyte colony-stimulating factor modulates the pulmonary host response to endotoxin in the absence and presence of acute ethanol intoxication. *J. Infect. Dis.* **179:**1441–1448.

42. Moses, J. M., Geschickter, E. H., and Ebert, R. H. (1968). The relationship of enhanced permeability to leukocyte mobilization in delayed inflammation. *Br. J. Exp. Pathol.* **49:**386–394.

43. Bagby, G. J., Zhang, P., Stoltz, D. A., and Nelson, S. (1998). Suppression of the granulocyte colony-stimulating factor response to *Escherichia coli* challenge by alcohol intoxication. *Alcohol Clin. Exp. Res.* **22:**1740–1745.

44. Zhang, P., Bagby, G. J., Boé, D. M., Zhong, Q., Schwarzenberger, P., Kolls, J. K. *et al.* (2002). Acute alcohol intoxication suppresses the CXC chemokine response during endotoxemia. *Alcohol Clin. Exp. Res.* **26:**65–73.

45. Liu, Y. K. (1980). Effects of alcohol on granulocytes and lymphocytes. *Semin. Hematol.* **17:**130–136.

46. McFarland, W. and Libre, E. P. (1963). Abnormal leukocyte response in alcoholism. *Ann. Intern. Med.* **59:**865–877.

47. Tisman, G. and Herbert, V. (1973). *In vitro* myelosuppression and immunosuppression by ethanol. *J. Clin. Invest.* **51:**1410–1414.

48. Imperia, P. S., Chikkappa, G., and Phillips, P. G. (1984). Mechanism of inhibition of granulopoiesis by ethanol. *Proc. Soc. Exp. Biol. Med.* **175:**219–225.

49. Hallengren, B. and Forsgren, A. (1978). Effect of alcohol on chemotaxis, adherence and phagocytosis of human polymorphonuclear leucocytes. *Acta Med. Scand.* **204:**43–48.

50. Tamura, D. Y., Moore, E. E., Partrick, D. A., Johnson, J. L., Offner, P. J., Harbedk, T. J. *et al.* (1998). Clinically relevant concentrations of ethanol attenuate primed neutrophil bactericidal activity. *J. Trauma* **44**:320–324.

51. Zhang, P., Nelson, S., Summer, W. R., and Spitzer, J. A. (1997). Acute ethanol intoxication suppresses the pulmonary inflammatory response in rats challenged with intrapulmonary endotoxin. *Alcohol Clin. Exp. Res.* **21**:773–778.

52. Sachs, C. W., Christensen, R. H., Pratt, P. S., and Lynn, W. S. (1990). Neutrophil elastase activity and superoxide production are diminished in neutrophils of alcoholics. *Am. Rev. Respir. Dis.* **141**:1249–1255.

53. Stoltz, D. A., Nelson, S., Kolls, J. K., Zhang, P., Bohm Jr., R. P., Murphey-Corb, M. *et al.* (2000). *In vitro* ethanol suppresses alveolar macrophage TNF-α during simian immunodeficiency virus infection. *Am. J. Res. Crit. Care Med.* **161**:135–140.

54. Zhang, Z., Cork, J., Ye, P., Lei, D., Schwarzenberger, P. O., Summer, W. R. *et al.* (2000). Inhibition of TNF-alpha processing and TACE-mediated ectodomain shedding by ethanol. *J. Leukoc. Biol.* **67**:856–862.

55. Kolls, J. K., Xie, J., Lei, D., Greenberg, S., Summer, W. R., and Nelson, S. (1995). Differential effects of *in vivo* ethanol on LPS-induced TNF and nitric oxide production in the lung. *Am. J. Physiol.* **268**:L991–L998.

56. Szabo, G., Mandrekar, P., Girouard, L., and Catalano, D. (1996). Regulation of human monocyte functions by acute ethanol treatment: Decreased tumor necrosis factor-alpha, interleukin-1 beta and elevated interleukin-10, and transforming growth factor-beta production. *Alcohol Clin. Exp. Res.* **20**:900–907.

57. Dannenberg, A. M. (1989). Immune mechanisms in the pathogenesis of pulmonary tuberculosis. *Rev. Infect. Dis.* **11**:S369–S378.

58. Mikszta, J. A., Waltenbaugh, C., and Kim, B. S. (1995). Impaired antigen presentation by splenocytes of ethanol-consuming C57BL/6 mice. *Alcohol* **12**:265–271.

59. Domiati-Saad, R. and Jerrells, T. R. (1993). The influence of age on blood alcohol levels and ethanol-associated immunosuppression in a murine model of ethanol consumption. *Alcohol Clin. Exp. Res.* **17**:382–388.

60. Chadha, K. C., Stadler, I., Albini, B., Nakeeb, S. M., and Thacore, H. R. (1991). Effect of alcohol on spleen cells and their functions in C57BL/6 mice. *Alcohol* **8**:481–485.

61. Shellito, J. E. (1998). Alcohol and host defense against pulmonary infection with *Pneumocystis carinii*. *Alcohol Clin. Exp. Res.* **22**:208S–211S.

62. Shellito, J. E. and Olariu, R. (1998). Alcohol decreases T-lymphocyte migration into lung tissue in response to *Pneumocystis carinii* and depletes T-lymphocyte numbers in the spleens of mice. *Alcohol Clin. Exp. Res.* **22**:658–663.

63. Bagasra, O., Kajdacys-Balla, A., Lischner, H. W., and Pomerantz, R. J. (1993). Alcohol intake increases human immunodeficiency virus type 1 replication in human peripheral blood mononuclear cells. *J. Infect. Dis.* **167**:789.

64. Bagasra, O., Backman, S.E. Jew, L., Tawadros, R., Cater, J., Boden, G. *et al.* (1996). Increased human immunodeficiency virus type 1 replication in human peripheral blood mononuclear cells induced by ethanol: Potential immunopathogenic mechanisms. *J. Infect. Dis.* **173**:550–558.

65. Saravolatz, L. D., Cerra, R. F., Pohlod, D. J. and Smereck, S. (1990). The effect of alcohol on HIV infection in vitro. *Prog. Clin. Biol. Res.* **325**:267–271.

66. Nair, M. P., Schwartz, S. A., Kronfol, Z. A., Heimer, E. P., Pottathil, R., and Greden, J. F. (1990). Immunoregulatory effects of alcohol on lymphocyte responses to human immunodeficiency virus proteins, *Prog. Clin. Biol. Res.* **325**:221.

67. Fong, I. W., Read, S., Wainberg, M. A., Chia, W. K., and Major, C. (1994). Alcoholism and rapid progression to AIDS after seroconversion. *Clin. Infect. Dis.* **19**:337–338.

68. Pol, S., Artru, P., Thepot, V., Berthelot, P., and Nalpas, B. (1996). Improvement of the CD4 cell count after alcohol withdrawal in HIV-positive alcoholic patients. *AIDS* **10**:1293–1294.

69. Staprans, S. I., Dailey, P. J., Rosenthal, A., Horton, C., Grant, R. M., Lerche, N. *et al.* (1999). Simian immunodeficiency virus disease course is predicted by the extent of virus replication during primary infection. *J. Virol.* **73:**4829–4839.

70. Smith, S. M., Holland, B., Russo, C., Dailey, P. J., Marx, P. A., and Conner, R. I. (1999). Retrospective analysis of viral load and SIV antibody responses in rhesus macaques infected with pathogenic SIV: Predictive value for disease progression. *AIDS Res. Human Retroviruses* **15:**1691–1701.

71. Nelson, S., Summer, W., Bagby, G., Nakamura, C., Stewart, L., Lipscomb, G. *et al.* (1991). Granulocyte colony-stimulating factor enhances pulmonary host defenses in normal and ethanol-treated rats. *J. Infect. Dis.* **164:**901–906.

72. Zhang, P., Bagby, G. J., Stoltz, D. A., Spitzer, J. A., Summer, W. R., and Nelson, S. (1997). Modulation of the lung host response by granulocyte colony-stimulating factor in rats challenged with intrapulmonary endotoxin. *Shock* **7:**193–199.

73. Stoltz, D. A., Zhang, P., Nelson, S., Bohm Jr., R. P., Murphey-Corb, M., and Bagby, G. J. (1999). Ethanol suppression of the functional state of polymorphonuclear leukocytes obtained from uninfected and simian immunodeficiency virus infected rhesus macaques. *Alcohol Clin. Exp. Res.* **23:**878–884.

74. Nelson, S., Belknap, S. M., Carlson, R. W., Dale, D., DeBoisblanc, B., Farkas, S. *et al.* (1998). A randomized controlled trial of filgrastim as an adjunct to antibiotics for treatment of hospitalized patients with community-acquired pneumonia. CAP Study Group. *J. Infect. Dis.* **178:**1075–1080.

75. Murray, H. W. (1994). Interferon-γ and host antimicrobial defense. *Am. J. Med.* **97:**459-467.

76. Beck, J. M., Liggitt, H. D., Brunette, E. N., Fuchs, H. J., Shellito, J. E., and Debs, R. J. (1991). Reduction in intensity of *Pneumocystis carinii* pneumonia in mice by aerosol administration of gamma interferon. *Infect. Immun.* **11:**3859–3862.

77. Skerrett, S. J. and Martin, T. R. (1992). Intratracheal interferon gamma augments pulmonary clearance of *Legionella pneumophila* in immunosuppressed rats. *Am. Rev. Respir. Dis.* **145:**A12.

78. Jaffe, H. A., Buhl, R., Mastrangeli, A., Holroyd, K. J., Saltini, C., Czerski, D. *et al.* (1991). Organ specific cytokine therapy. Local activation of mononuclear phagocytes by delivery of an aerosol of recombinant interferon-gamma to the human lung. *J. Clin. Invest.* **88:**279-302.

79. Kolls, J. K., Lei, D., Stoltz, D., Zhang, P., Schwarzenberger, P. O., Ye, P. *et al.* (1998). Adenoviral-mediated interferon-gamma gene therapy augments pulmonary host defense ethanol-treated rats. *Alcohol Clin. Exp. Res.* **22:**157–162.

80. Lei, D., Lancaster Jr., J. R., Joshi, M. S., Nelson, S., Stoltz, D., Bagby, G. J. *et al.* (1997). Activation of alveolar macrophages and lung host defenses using transfer of the interferon-gamma gene. *Am. J. Physiol.* **272:**L852–L859.

81. Zhang, P., Quinton, L. J., Bagby, G. J., Summer, W. R., and Nelson, S. (2003). Interferon-γ enhances the pulmonary CXC chemokine response to intratracheal lipopolysaccharide challenge. *J. Infect. Dis.* **187:**62–69.

82. Gallin, J. I., Farber, J. M., Holland, S. M. and Nutman, T. B. (1995). Interferon-γ in the management of infectious diseases. *Ann. Intern. Med.* **123:**216–224.

83. Condos, R., Rom, W. N., and Schluger, N. W. (1997). Treatment of multidrug-resistant pulmonary tuberculosis with interferon-gamma via aerosol. *Lancet* **349:**1513–1515.

Index